Springer Theses

Recognizing Outstanding Ph.D. Research

Aims and Scope

The series "Springer Theses" brings together a selection of the very best Ph.D. theses from around the world and across the physical sciences. Nominated and endorsed by two recognized specialists, each published volume has been selected for its scientific excellence and the high impact of its contents for the pertinent field of research. For greater accessibility to non-specialists, the published versions include an extended introduction, as well as a foreword by the student's supervisor explaining the special relevance of the work for the field. As a whole, the series will provide a valuable resource both for newcomers to the research fields described, and for other scientists seeking detailed background information on special questions. Finally, it provides an accredited documentation of the valuable contributions made by today's younger generation of scientists.

Theses are accepted into the series by invited nomination only and must fulfill all of the following criteria

- They must be written in good English.
- The topic should fall within the confines of Chemistry, Physics, Earth Sciences, Engineering and related interdisciplinary fields such as Materials, Nanoscience, Chemical Engineering, Complex Systems and Biophysics.
- The work reported in the thesis must represent a significant scientific advance.
- If the thesis includes previously published material, permission to reproduce this must be gained from the respective copyright holder.
- They must have been examined and passed during the 12 months prior to nomination.
- Each thesis should include a foreword by the supervisor outlining the significance of its content.
- The theses should have a clearly defined structure including an introduction accessible to scientists not expert in that particular field.

More information about this series at http://www.springer.com/series/8790

Michael Werner Zürch

High-Resolution Extreme Ultraviolet Microscopy

Imaging of Artificial and Biological Specimens with Laser-Driven Ultrafast XUV Sources

Doctoral Thesis accepted by
Friedrich Schiller University of Jena, Germany

Author
Dr. Michael Werner Zürch
Institute of Optics and Quantum Electronics
Friedrich Schiller University of Jena
Jena
Germany

Supervisor
Prof. Christian Spielmann
Institute of Optics and Quantum Electronics
Friedrich Schiller University of Jena
Jena
Germany

ISSN 2190-5053 ISSN 2190-5061 (electronic)
ISBN 978-3-319-38565-5 ISBN 978-3-319-12388-2 (eBook)
DOI 10.1007/978-3-319-12388-2

Springer Cham Heidelberg New York Dordrecht London

Softcover reprint of the hardcover 1st edition 2015

Printed on acid-free paper

Springer is part of Springer Science+Business Media (www.springer.com)

Für Sara

Supervisor's Foreword

Microscopy is in many fields of science the most important tool to investigate and/or classify small structures. In the nineteenth century, the famous scientist Ernst Abbe, professor at the University of Jena, derived an equation to estimate the highest possible resolution in microscopy. The Abbe limit states that even for perfect optical systems (with the highest possible numerical aperture of the lens system of the order of one) the smallest resolvable feature is of the order of the wavelength of the illuminating light source. Following Abbe's prediction, many interesting elements in nano-optics and life science such as bacteria and viruses cannot be resolved with optical microscopes. A very promising route to overcome this limitation is employing light sources in the extreme ultraviolet or soft X-ray range. Unfortunately, optical components in this wavelength range suffer either from a low transmission or a very small numerical aperture preventing high-resolution microscopy with lab sources. In the present book, Michael Zürch presents novel concepts and realizations for extreme ultraviolet microscopy with lab sources. Illuminating the sample with coherent short wavelength radiation enables a reconstruction of the object from the scattered light recorded, e.g., with a CCD detector. In such a setup the task of the imaging optics is carried out by a computer algorithm and the whole approach is known in the literature as lensless imaging or coherent diffraction imaging (CDI).

After a short introduction, Michael Zürch provides an overview of the fundamentals of lensless imaging with short wavelength lab sources. The description of the laser-driven high harmonic generation (HHG) source is short but very concise. More expanded is the discussion about the basics, limitations, and different realizations of lensless imaging. Highly appreciated is his detailed discussion on how to distinguish between coherent diffraction imaging and holography, to be more precise, in-line holography. He illustrates the most important features for each of these two techniques and discusses experimental peculiarities that occurred. The chapter is concluded with a detailed and carefully investigated overview of the recent publications in this area, including state-of-the-art reconstruction algorithms. It is very helpful to get acquainted with this topic, because it is not just a listing of recent publications, but a discussion and comparison of many distinct results. In the third chapter, Michael Zürch presents the experimental setups used in his work. He reports on the application of two different laser sources for HHG, namely a more widely used Ti:sapphire system and a newly developed Yb-doped fiber laser setup. The latter has the advantage of very high average power, which will help to reduce the exposure time significantly, paving the way for real-time microscopy with a short wavelength radiation. Further innovative approaches to spectrally select and steer the short wavelength radiation are briefly explained.

After the description of the experimental setup, many interesting results are presented: The distribution and size of nanoparticles on a thin membrane has been evaluated with in-line holography. Illuminating the samples with coherent light enabled a reconstruction of the 3D objects from the recorded hologram. Nanostructured transmission samples have been imaged with the narrowband HHG radiation from the fiber laser source. In agreement with the predictions made in the previous chapters, the CDI reconstruction showed the highest possible resolution as suggested by the Abbe limit. These results represent a new record for the resolution obtained with laser-driven short wavelength sources. Equally interesting besides these optimizations and best results for transmission geometry are the results for the reflection geometry. As shown in the present work, this imaging geometry requires many additional careful corrections of the recorded image prior to the reconstruction in order to obtain meaningful results. Here Michael Zürch did pioneering work, and succeeded for the first time in obtaining high-resolution imaging in reflection geometry. This approach is well suited for real-world applications, because, e.g., biological samples need only be pipetted on a thick substrate and not prepared on a very thin and fragile membrane as necessary for the transmission geometry. In the following proof-of principle experiment he demonstrated that the recorded diffraction pattern contains enough information to classify different cancer cells. Based on these experiments, a new fast and reliable method to classify cells will be in reach, and has the potential to be used for routine clinical diagnostics. In the final chapter, Michael Zürch put forward a new approach for high-resolution CDI: So far in all imaging applications the illuminating radiation had either a flat or spherical wave front. Now he discusses the possibility of a beam with a helical wave front, which is also known in the literature as an optical vortex. The additional phase modulation results in a more pronounced diffraction pattern holding the promise for improved imaging, as shown by simulation. Prior to imaging applications with a short wavelength vortex beam, such a beam must be demonstrated experimentally. HHG with a visible laser beam carrying a helical phase-front is a promising route toward short wavelength vortex beams. The first successful completion of the generation experiment attracted a lot of interest in the community and has been published in a highly ranked journal.

In order to realize high-resolution microscopy with short wavelengths it was necessary to combine state-of-the-art laser systems with concepts of classical and nonlinear optics, including modern image reconstruction algorithms. Michael Zürch did it in an extremely convincing and efficient way, as can be seen in the present book. Additionally, the work has been published in many internationally journals, presented at many international conferences, and two patents have been filed. Needless to say the contributions attracted a lot of attention. Noteworthy is that his work on the classification of cancer cells has been awarded as the best contribution at a large international conference on medical imaging. Summing up, it was really a great pleasure to supervise such a talented and focused young researcher, and I am confident that he will definitely make a great career in science. Finally, I hope you enjoy reading this book as much as I did.

Jena, August 2014 Prof. Christian Spielmann

Abstract

This thesis studies different realizations of imaging of artificial and biological specimens at high spatial resolution by using coherent extreme ultraviolet (XUV) radiation. These so-called *lensless imaging* techniques dispense with using imaging optical elements and instead directly detect the diffraction pattern of the object, which is subsequently reconstructed by means of iterative algorithms on a computer. The light source used is based on high harmonic generation driven by an ultrafast laser in a rare gas. The techniques presented bring this high-resolution method, which has been mostly used at large-scale facilities such as synchrotrons and free-electron lasers, to the lab scale.

Two different lensless imaging systems are presented in this work. One uses a grating to monochromatize the radiation produced, while the other uses dielectric coated mirrors for that purpose. The latter produces a clean and tight focus to be used to illuminate the object. The grating-based system is, for example, useful to image larger samples with different wavelengths.

If a tiny pinhole is used in front of the sample, one produces a spherical illumination wave, which gives rise to holographic imaging. It was shown in this work that this can be used to image complexly shaped and extended non-isolated objects in great detail. The objects used were specifically prepared in order to mimic biological specimens. Furthermore, it was possible to obtain three-dimensional information about the sample by either analyzing the phase or the absorption. Another benefit demonstrated is the large material contrast in the XUV, although the spatial resolution presented is in the range of good optical microscopes.

Moreover, lensless imaging at high numerical aperture was investigated in this work. A spatial resolution on the order of a few tens of nanometers was achieved, which corresponds to less than two wavelengths of the illumination light used. Depending on the method to determine the resolution achieved there are indications that the resolution achieved is below one wavelength and thus close to the Abbe limit. This shows that compact ultrafast laser-driven sources offer excellent coherence properties and allow imaging at the diffraction limit.

It was also shown in this thesis that objects can be imaged in reflection geometry. Up to now lensless imaging was almost exclusively used in transmission geometry with transparent samples or apertures. The reflection geometry is for example important when optical nanostructures or biological specimens are examined. Especially for the latter this is important, because they can be pipetted directly on a substrate, which usually rules out transmission geometry, because substrates—except fragile ultrathin membranes—are typically opaque. Furthermore, these specimens are opaque, but reflect well, themselves; hence, the surface can be examined in reflection geometry. From the complex-valued object plane reconstruction one can draw further information about the specimen, e.g., the height profile.

By means of a method developed in the work presented it appears possible to distinguish different cell expressions directly from the XUV diffraction pattern. The characterization is solely based on outer shape features of the specimen. In a proof-of-principle experiment this method was applied to distinguish between two different human breast cancer cell expressions for which several diffraction patterns were recorded.

Furthermore, it was shown for the first time that so-called *optical vortex* beams can directly be converted from the infrared into the XUV spectral domain by means of high harmonic generation. It was shown that the observed XUV vortex beams have properties deviating from theoretical expectations, for which explanations were already found in other experiments, leaving room for further experiments on that matter. These XUV vortex beams might soon gain special importance in the spectroscopy of forbidden transitions or for high-resolution imaging below the Abbe limit.

Acknowledgments

In this section I would like to express my thanks to those who made this work possible, because a scientific work of this extent is the result of the interaction of many people. This is by no means a complete list of people who made substantial contributions, thus my thanks go to all who helped me to accomplish this work and for making it so enjoyable.

In particular, I wish to thank my doctoral advisor Prof. C. Spielmann for taking me into his group, for continuous generous support in bringing my ideas to reality, and last but not least, for this interesting project that brought me so much further than I would have ever expected.

I am grateful to our talented lab engineers, W. Ziegler and M. Damm, for their extensive support in constructing and assembling experiments and so much more. These experiments would not have been possible without the outstanding support of our scientific workshops. I am especially thankful to B. Klumbies, R. Bark, and P. Engelhardt for their help in particular when things had to be done *quickly*, and for their devotion to building perfect experimental equipment.

Moreover, I am thankful to the colleagues of the Quantum Electronics department and the IOQ for the helpful inspiring discussions, for the great working atmosphere, and for so much more. The contributions of the numerous undergraduate students are much appreciated.

Further, I would like to express my thanks to many colleagues, whose contributions were invaluable to this work and could not be duly recognized without another 100 pages: Dr. J. Rothhardt, Dr. S. Hädrich, S. Wolf, H.-J. Hempel, Prof. Dr. H. Gross, Dr. I. Uschmann, R. Kuth, S. Förtsch, Prof. Dr. K. Pachmann, T. Weber, C. Kern, Dr. M. Schnell, Prof. Dr. A. Dreischuh, A. Guggenmos… many thanks!

For proofreading of the manuscript and helpful suggestions I am grateful to: Dr. D. Kartashov, Dr. M. Kanka, Dr. S. Hädrich, and P. Hansinger. Many thanks to Dr. O. Bock and N. Cheadle for improving the layout and language.

Further, the reviews and comments for improving the manuscript from Prof. Dr. A. Szameit and Prof. Dr. S.R. Leone are much appreciated.

Extraordinary thanks go to my family for their enduring loving support during the studies and my doctorate. Their trust in my abilities brought me all the way. The greatest thanks that can hardly be put into words go to my best friend and loving wife Sara, who was always there when I needed her and who thoroughly supported my plans. Thank you!

Contents

1 Preamble . . . 1
References . . . 3

2 Introduction and Fundamental Theory . . . 5
2.1 High Harmonic Generation in Gases . . . 5
2.1.1 Ionization . . . 7
2.1.2 Acceleration . . . 8
2.1.3 Recombination . . . 9
2.2 Coherent Diffraction Imaging Theory . . . 10
2.2.1 Coherence Properties of a Light Field . . . 10
2.2.2 Diffraction of Light Waves at Matter . . . 11
2.2.3 The Phase Problem . . . 16
2.2.4 Geometric Considerations . . . 19
2.3 Iterative Phase Retrieval Methods . . . 21
2.4 Digital In-line Holography . . . 25
2.5 CDI State-of-the-Art and Chapter Summary . . . 32
References . . . 33

3 Experimental Setup . . . 41
3.1 Realization and Characteristics of the HHG Source . . . 41
3.1.1 Ultrafast Ti:Sa Laser Based HHG Source . . . 41
3.1.2 Fiber CPA Driven HHG Source . . . 44
3.2 Table-Top CDI Schemes . . . 45
3.2.1 Grating Based CDI Setup . . . 45
3.2.2 Dielectric Mirror Based Setup . . . 48
3.2.3 Enhancing the Dynamic Range of the Diffraction Pattern . . . 50
3.3 XUV Digital In-line Holography Realization . . . 53

3.4 Preparation of the Experimental Data 54
3.4.1 Intensity Normalization and Curvature Correction..... 54
3.4.2 Reflection Geometry.......................... 58
References 61

4 Lensless Imaging Results 65
4.1 Digital In-line Holography of Cell-Like Non-periodic Specimens.. 65
4.1.1 Cell-Like Sample Preparation for DIH XUV Experiments 66
4.1.2 DIH Results Using Coherent XUV Radiation........ 68
4.1.3 Summary of the DIH XUV Experiments........... 73
4.2 Transmission CDI at the Abbe Limit with a High Numerical Aperture 76
4.2.1 High NA Reconstruction of a Pinhole Aperture 76
4.2.2 High NA Reconstruction of a Complex-Shaped Aperture 78
4.2.3 Summary of the Transmission CDI Experiments 80
4.3 CDI of an Artificial Non-periodic Specimen in Reflection Geometry 81
4.4 CDI Based Fast Classification of Breast Cancer Cells........ 85
References 91

5 Optical Vortices in the XUV............................ 95
5.1 Singular Light Beams—Optical Vortices................. 95
5.2 Experimental Setup for Generating XUV Vortex Beams 97
5.3 First Demonstration of Optical Vortices in the XUV......... 100
5.4 Application of XUV Vortex Beams to High Resolution Lensless Imaging 104
References 106

6 Summary and Outlook.............................. 109
References 114

Appendix A: XUV Mirror Ray Trace and Aberration Minimized Setup.............................. 117

Appendix B: Simulations of Fringe Patterns for Different Topological Charges in XUV Vortex Beams.......... 121

Curriculum Vitae 123

Abbreviations and Symbols

$\hbar$	Planck's constant
λ	Wavelength of the light
$\mathbf{q}$	Momentum transfer vector, which describes the momentum change in an elastic scattering experiment.
$\mathcal{F}$, FT	*Fourier Transform*
Δr	Spatial resolution, either used as theoretical limit or experimentally achieved resolution, depending on context.
ξ_l	Longitudinal coherence length, also often referred to as temporal coherence.
ξ_t	Transverse coherence length, also often referred to as spatial coherence.
c	Speed of light
e	Charge of the electron
f	Focal length of a lens or a focusing mirror
I	Intensity, typically measured in W/cm^2
O	Linear oversampling degree of a two dimensional diffraction pattern.
S	Support function, is a step function which is one inside the object and zero outside the object.
CCD	*Charge Coupled Device*, a two-dimensional detector to measure spatial intensity distributions. For all experimental data presented in this thesis an XUV sensitive Andor iKonL (2048 × 2048 pixels, $(13.5 \times 13.5)\ \mu m^2$ pixel size) CCD camera was used.
CDI	*Coherent Diffraction Imaging*, a technique that allows imaging without using optical elements, e.g., lenses.
CPA	*Chirped Pulse Amplification*, scheme to amplify ultrashort laser pulses by first temporally stretching them (chirping), amplifying them, and subsequently compressing them again.
DIH	*Digital In-line Holography*, a holography technique where the in-line hologram is recorded by a CCD.
eV	*Electronvolt*, is a unit of energy approximately equal to 1.609×10^{-19} J. Wavelength conversion: $E(\mathrm{eV}) = \frac{1239.84\,\mathrm{eV\,nm}}{\lambda(\mathrm{nm})}$.

FCPA	*Fiber CPA*, a laser system driven by an ultrafast fiber laser using the CPA scheme.
FEL	*Free-Electron Laser*, a laser producing coherent X-ray radiation.
FFT	*Fast Fourier Transforms*, a numerical implementation for the Fourier transform.
FOV	*Field of View*, the accessible extent of the image in object plane.
FWHM	*Full Width Half Maximum*, i.e., the width of the signal at half of its maximum on the respective axis.
GHIO	*Guided Hybrid Input–Output* algorithm, a technique to average independent reconstruction runs after a certain amount of iterations to find a better global solution.
HDR	*High Dynamic Range*, in this thesis HDR denotes a measured diffraction pattern stitched together from single patterns recorded with different exposure times in order to get a diffraction pattern with a dynamic range exceeding the dynamic range of the CCD.
HHG	*High Harmonic Generation*, a highly nonlinear process that coherently converts laser light to a spectrum of harmonics of the laser frequency typically expanding over many harmonic orders.
HIO	*Hybrid Input–Output* algorithm, the working horse in iterative phase retrieval.
NA	*Numerical Aperture*, is the sine of the half opening angle of a imaging system times the index of refraction.
OV	*Optical Vortex*, is a special beam shape, where the light beam carries a screw-like phase profile and doughnut-like intensity profile. They are also termed *singular light beams*, due to their phase singularity.
PBS	*Phosphate buffered saline*, a solution that imitates the pH value of the human organism.
PCR	*Polymerase Chain Reaction*, a method for DNA analysis based on multiplication of the DNA strains.
PRTF	*Phase Retrieval Transfer Function*, a function that analytically determines the quality of a reconstruction and gives rise to the achieved resolution.
RAAR	*Relaxed Averaged Alternating Reflection* algorithm, an improved version of the HIO.
SEM	*Scanning Electron Microscope*, a high-resolution scanning microscope using electrons as probe.
SLM	*Spatial Light Modulator*, a device that is addressed like a computer display and allows pixelwise phase manipulation of a laser beam.
TC	*Topological Charge*, a measure of how often the phase of an OV beam is wrapped modulo 2π in one turn across the beam's azimuthal coordinate ϕ.
Ti:Sa	*Titanium-doped Sapphire laser*, a solid state ultrafast laser based on a Al_2O_3 crystal doped with Ti^{3+}.
XUV	*Extreme Ultraviolet*, part of the electromagnetic spectrum with wavelengths ranging from 124 nm down to about 10 nm.

Chapter 1
Preamble

It must have been an astonishing experience for people when they took a look into ZACHARIAS JANSSEN's[1] light microscope, which he invented in the late 17th century. Ever since those days the quest to make smaller details visible has been pursued, and continues unabated today. While optical microscopy was one of the driving motors for science over the centuries, tremendous progress was made in the 20th century with the advent of fluorescence microscopy and scanning microscopy techniques. With the latter one can nowadays image structures down to the atomic level. However, most of these specialized microscopy techniques require special sample preparation or are limited to specific materials. Thus optical imaging microscopy can still be considered one of the most important microscopy techniques. The fundamental limitation for optical imaging microscopes was found in the 1870s by ERNST ABBE.[2] His resolution limit

$$d = \frac{\lambda}{2n \sin \alpha} \tag{1.1}$$

basically describes that the smallest resolvable structure d is limited to half of the wavelength λ of the illuminating light for a given index of refraction n and a half opening angle α of the objective. Since the index of refraction is about one and the sine of the opening angle cannot be greater than one, the fundamental limitation is induced by the wavelength of the light. For that reason efforts were undertaken to use X-ray tubes and Fresnel zone plates as lenses in order to construct X-ray microscopes. The resolution of these devices, however, was far away from being limited by the wavelength, since Fresnel zone plates cannot be fabricated with sufficiently small details on a larger area in order to allow for a high numerical aperture. Nevertheless these incoherent X-ray sources are still very important for diffraction studies on crystalline material.

---

[1] Zacharias Janssen, *1588 in Den Haag, †1631 in Amsterdam.

[2] Ernst Abbe, *1840 in Eisenach, †1905 in Jena.

M.W. Zürch, *High-Resolution Extreme Ultraviolet Microscopy*,
Springer Theses, DOI 10.1007/978-3-319-12388-2_1

With the advent of coherent X-ray sources, such as synchrotrons (small coherence) and more recently the highly coherent free-electron lasers (FEL) things changed again. Using the coherent nature of the light one could image structures using their continuous far-field diffraction pattern without being limited in resolution by optical devices such as zone plates, by the high absorption of solid matter or the need for crystalline nature of the object under investigation. Coherent diffraction imaging (CDI), which is the name of this field, has since then become a versatile direct imaging technique with a high impact [1]. For instance, the determination of a protein using lensless imaging [2] at the free-electron laser in Stanford (SLAC) was chosen as one of the top ten scientific breakthroughs in 2012 in the *Science* magazine. Another remarkable example of the capabilities was the 3D tomographic measurement of a yeast spore cell with a resolution down to a few nanometers [3].

However, for broad application of CDI in life science the need for a FEL or next-generation synchrotron is limiting. An alternative light source is based on high harmonic generation (HHG), which allows to convert visible ultrafast laser light into coherent extreme ultraviolet (XUV) light at the lab scale. These laser-driven XUV sources are compact and feature a highly coherent laser-like light beam which makes them perfect for coherent imaging techniques. Since the first demonstration [4] of CDI using HHG this field has experienced tremendous progress [5, 6].

Another approach to achieve high resolution imaging, often termed *super-resolution imaging*, well below the Abbe limit, uses structured illumination [7]. In this work an approach towards structured illumination imaging in the XUV will be demonstrated by converting a so-called optical vortex (OV) beam into the XUV by HHG [8]. Besides possible enhancements for imaging, a great benefit for spectroscopy could be drawn from having such shaped light beams available at shorter wavelengths.

In this thesis several advances regarding table-top HHG and its application to imaging, including OV beam generation in the XUV, will be demonstrated. In Chap. 2 some fundamentals regarding HHG, coherent imaging techniques such as CDI and digital in-line holography will be discussed. Moreover, the iterative algorithms used to reconstruct the images will be introduced in that chapter. In Chap. 3 the experimental setup that was used for different experiments is described, including different schemes for coherent imaging, i.e. CDI and digital in-line holography. An important part of the work will deal with reflection geometry CDI [9] which is of special importance for biological imaging. In the subsequent Chap. 4 the experimental findings regarding imaging will be presented. The latest results of CDI measurements of a high numerical and an achieved resolution of the order of a wavelength will be presented. Digital in-line holography images taken from cell-like non-periodic specimens will be shown. Moreover, the results consist of a CDI image taken from a non-periodic artificial specimen in reflection geometry, and, finally, the demonstration of a novel fast CDI based classification technique for breast cancer cells. The results obtained for OV generation in the XUV, together with an outlook on possible imaging schemes that could possibly contribute towards nanometer resolution, will be presented in Chap. 5. The last chapter summarizes the main results and provides an outlook for future experiments.

References

1. Chapman, H.N., Nugent, K.A.: Coherent lensless X-ray imaging. Nat. Photon. **4**(12), 833–839 (2010)
2. Boutet, S., Lomb, L., Williams, G.J., Barends, T.R.M., Aquila, A., Doak, R.B., Weierstall, U., DePonte, D.P., Steinbrener, J., Shoeman, R.L., Messerschmidt, M., Barty, A., White, T.A., Kassemeyer, S., Kirian, R.A., Seibert, M.M., Montanez, P.A., Kenney, C., Herbst, R., Hart, P., Pines, J., Haller, G., Gruner, S.M., Philipp, H.T., Tate, M.W., Hromalik, M., Koerner, L.J., van Bakel, N., Morse, J., Ghonsalves, W., Arnlund, D., Bogan, M.J., Caleman, C., Fromme, R., Hampton, C.Y., Hunter, M.S., Johansson, L.C., Katona, G., Kupitz, C., Liang, M.N., Martin, A.V., Nass, K., Redecke, L., Stellato, F., Timneanu, N., Wang, D.J., Zatsepin, N.A., Schafer, D., Defever, J., Neutze, R., Fromme, P., Spence, J.C.H., Chapman, H.N., Schlichting, I.: High-resolution protein structure determination by serial femtosecond crystallography. Science **337**(6092), 362–364 (2012)
3. Jiang, H., Song, C., Chen, C.C., Xu, R., Raines, K.S., Fahimian, B.P., Lu, C.H., Lee, T.K., Nakashima, A., Urano, J., Ishikawa, T., Tamanoi, F., Miao, J.: Quantitative 3D imaging of whole, unstained cells by using X-ray diffraction microscopy. Proc. Natl. Acad. Sci. USA **107**(25), 11234–11239 (2010)
4. Sandberg, R.L., Paul, A., Raymondson, D.A., Haedrich, S., Gaudiosi, D.M., Holtsnider, J., Tobey, R.I., Cohen, O., Murnane, M.M., Kapteyn, H.C., Song, C.G., Miao, J.W., Liu, Y.W., Salmassi, F.: Lensless diffractive imaging using tabletop coherent high-harmonic soft-X-ray beams. Phys. Rev. Lett. **99**(9), 098103 (2007)
5. Miao, J.W., Sandberg, R.L., Song, C.Y.: Coherent X-ray diffraction imaging. IEEE J. Sel. Top. Quant. Electron. **18**(1), 399–410 (2012)
6. Sandberg, R.L., Huang, Z.F., Xu, R., Rodriguez, J.A., Miao, J.W.: Studies of materials at the nanometer scale using coherent X-ray diffraction imaging. JOM **65**(9), 1208–1220 (2013)
7. Dan, D., Lei, M., Yao, B., Wang, W., Winterhalder, M., Zumbusch, A., Qi, Y., Xia, L., Yan, S., Yang, Y., Gao, P., Ye, T., Zhao, W.: DMD-based LED-illumination super-resolution and optical sectioning microscopy. Sci. Rep. **3**, 1116 (2013)
8. Zuerch, M., Kern, C., Hansinger, P., Dreischuh, A., Spielmann, C.: Strong-field physics with singular light beams. Nat. Phys. **8**(10), 743–746 (2012)
9. Zuerch, M., Kern, C., Spielmann, C.: XUV coherent diffraction imaging in reflection geometry with low numerical aperture. Opt. Express **21**(18), 21131–21147 (2013)

Chapter 2
Introduction and Fundamental Theory

In this chapter the fundamentals involved in coherent imaging with laser generated XUV light are presented. The first Sect. 2.1 introduces high harmonic generation, which is the process that is used throughout this thesis to generate coherent XUV light. In Sect. 2.2 some imaging theory and a mathematical description for the diffraction of light by matter will be introduced. A major issue in CDI is the so-called *phase problem*, which arises from the fact that one can only measure intensities with a physical detector in an experiment. For the reconstruction of an object, however, the phase of the light field must be known. If sufficient care to some geometric constraints, as presented in Sect. 2.2.4, is taken, one can retrieve the phase of the light field by means of iterative algorithms, which are introduced in Sect. 2.3. Another approach for coherent imaging is digital in-line holography, which is a technique where the phase of the light field is already encoded in the fringes measured in the far-field with a detector. The fundamentals about digital in-line holography are presented in Sect. 2.4. A chapter summary and view at other modern imaging techniques, which are not used in this thesis, will conclude this chapter

2.1 High Harmonic Generation in Gases

In this section the basic nonlinear process to generate short wavelength light will be introduced. Therefore, first an electro-magnetic wave impinging on a material is considered as the interaction between light and matter is discussed. The feedback from the material to the electric field $(\mathbf{E}(\mathbf{r}, t))$[1] is expressed by a linear dependence of the polarisation $\mathbf{P}$ on the electric field connected by the susceptibility χ. This linear behavior corresponds to bound electrons moving in the Coulomb potential of the atoms, and experiencing a small perturbation of their motion by the laser field. This small perturbation can be treated as a small amplitude electron bouncing in a

[1] Throughout this thesis vectors and vectorial quantities will be denoted by a boldface typeset, e.g. $\mathbf{r}$. The modulus of $\mathbf{r}$ is denoted simply as r.

M.W. Zürch, *High-Resolution Extreme Ultraviolet Microscopy*,
Springer Theses, DOI 10.1007/978-3-319-12388-2_2

harmonic potential approximating the atomic or molecular potential in the vicinity of the stationary orbit. However, this approximation is not valid for higher intensities and thus the relation between the polarization and the electric field turns from linear to a power series in χ [1]

$$\mathbf{P}(\mathbf{r},t)_{\rm NL} \propto \chi^{(1)}(\omega)\mathbf{E}(\mathbf{r},t) + \chi^{(2)}(\omega)\mathbf{E}^2(\mathbf{r},t) + \chi^{(3)}(\omega)\mathbf{E}^3(\mathbf{r},t) + \cdots, \quad (2.1)$$

where $\chi^{(n)}$ represents the n-th order susceptibilities. For non-centrosymmetric media, all even terms in this series are equal to zero. The nonlinear response of the polarization results in a zoo of effects that kick in once a certain intensity, depending on material parameters, is reached. The coefficients $\chi^{(n)}$ can be linked to a specific nonlinear effect, e.g. harmonic generation. More details are discussed in [2].

The most used nonlinear effect is frequency conversion, which can be anything from multiple harmonic generation to complicated frequency mixing [1]. However, due to absorption and usually decreasing $\chi^{(n)}$ for increasing n for typical materials, the generation of harmonics much beyond the third order is not efficient.

Expression 2.1 is the basis for conventional perturbative nonlinear optics. It is obvious that the nonlinear motion of bound electrons is not able to provide harmonics with the energy of quanta larger than the ionization potential of the atom or molecule. To overcome this limitation and to enable frequency conversion to the XUV and X-ray spectral range, a new mechanism inevitably involving the ionization of the medium in a high intensity laser field should be proposed. Such a mechanism was proposed in [3, 4] and is called *high-order harmonic generation* (HHG) in gases. This highly nonlinear process allows to generate coherent light with photon energies up to the kiloelectronvolt level [5, 6]. The principal structure of an HHG spectrum is depicted in Fig. 2.1. The intensities of the lower order odd harmonics only, with any energy of quantas up to the ionization potential of the atom, follow the perturbative regime predicted by Eq. 2.1. These lower order harmonics are followed by the so called plateau, where the yield of the harmonics is about constant over a large wavelength span. A cut-off at the high energetic end of the spectrum relates to the shortest produced wavelengths.

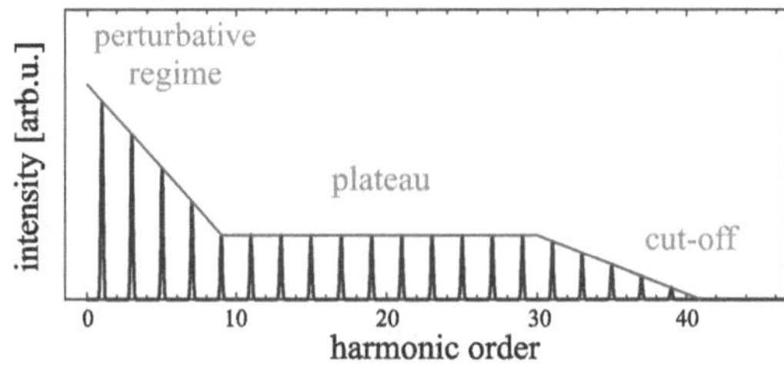

Fig. 2.1 Schematic HHG spectrum. The intensity of the lower order harmonics follows a perturbative regime, which is followed by a plateau featuring about equally intense harmonics. The cut-off marks the highest photon energies generated

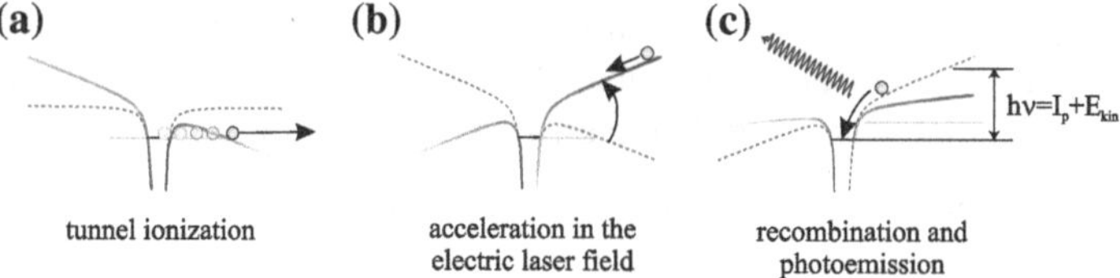

Fig. 2.2 The three-step-model in brief: **a** An electron tunnels out of the parent ion due to the Coulomb potential (*dotted red line*) of the atom being deformed (*solid red line*) by the electric field of the laser. **b** The electron accelerates in the laser field and gains kinetic energy. **c** The electron recombines with the parent ion and a photon with a energy $h\nu$ equivalent to the electrons kinetic energy $E_{\rm kin}$ and the ionization potential I_p of the ion is emitted

In the early 1990s Krause et al. [7] and Corkum [8] developed a simple model that is able to describe the plateau and cut-off structure for a single-atom using a semi-classical approach. The so-called three-step-model consists of three parts: ionization, acceleration, and recombination of an electron, as depicted in Fig. 2.2.

2.1.1 Ionization

In order to estimate the electric field that is needed to significantly modify the Coulomb potential of an atom, one can calculate the electric field of a bound electron in a hydrogen atom. This is done by using Bohr's atomic model and the Bohr radius a_0 to calculate the electric field as

$$E_a = \frac{e}{4\pi\epsilon_0 a_0^2} = 5 \times 10^{11}\,\mathrm{V/m}. \tag{2.2}$$

From this one gets an estimate of the intensity of a laser field having an equivalent electric field

$$I_a = \frac{1}{2}\epsilon_0 c E_a^2 = 3.51 \times 10^{16}\,\mathrm{W/cm^2}. \tag{2.3}$$

Such intensities are easily reachable with modern ultrafast lasers. Using the single active electron (SAE) approximation [9], one can write the combined potential experienced by the single electron and the laser field as

$$V(\mathbf{r}, t) = -\frac{e^2}{4\pi\epsilon_0|\mathbf{r}|} + e\mathbf{E}(t)\mathbf{r}. \tag{2.4}$$

Obviously, a significant modification of the electron motion in the atom can only take place if the intensity of the laser field is high enough, i.e. on the order given in Eq. 2.3.

According to Eq. 2.4, the combined potential features a time-dependent barrier. The width of this barrier r_B can be estimated as $er_B E \sim I_p$, where I_p is the ionization potential of the atom. Using the characteristic velocity of the electron motion in the atom $v_a \sim \sqrt{\frac{2I_p}{m_e}}$, one can introduce a characteristic tunneling time $\tau \sim \frac{r_B}{v_a} \sim \sqrt{\frac{2m_e I_p}{e^2 E^2}}$. The regime of ionization is determined by the ratio of the tunneling time to the optical cycle. A corresponding parameter, known as the *Keldysh parameter*, to distinguish these regimes was introduced by Keldysh [10] and is defined as

$$\gamma = \tau\omega = \sqrt{\frac{I_p}{2U_p}}, \tag{2.5}$$

where ω is the laser frequency and $U_p = \frac{e^2 E_0^2}{4m_e\omega^2}$ is the ponderomotive potential, which is the mean kinetic energy of a free electron oscillating in the monochromatic laser field with an amplitude E_0. When $\gamma \gg 1$, which corresponds to relatively low laser intensities or short laser wavelengths, the electron does not have sufficient time to tunnel through the potential barrier within an optical cycle. Instead, ionization proceeds via absorption of multiple optical photons. This is the so-called *multiphoton ionization* regime. It is typical for high intensity laser pulses in the visible or UV spectral range. In the opposite case $\gamma \ll 1$, ionization occurs on the sub-cycle time scale due to electron tunneling through the potential barrier. For the near-infrared laser wavelength 800 nm, as used in the experiments in this thesis, ionization occurs in a mixed regime, since $\gamma \sim 1$.

2.1.2 Acceleration

The generation of a free electron is followed by acceleration (Fig. 2.2b) in the laser field $E = E_0 \cos(\omega t + \phi)$. Since ionization occurs in the electric field of the laser pulse, which is comparable to the characteristic atomic field (Eq. 2.2), the Coulomb attraction force from the ion on the free electron is much weaker than the electric force of the laser pulse. That is why the influence of the ion on the electron motion can be neglected and analysis can be done for trajectories of a free electron in the monochromatic electric field. Hence, one can calculate the movement of the electron that is born at a phase ϕ of the laser field, assuming the electron being at rest $v_0 = 0$ after tunneling, by

$$v(t) = \int_0^t -\frac{e}{m_e}E(t)\mathrm{d}t = -\frac{E_0 e}{m\omega}\{\sin(\omega t + \phi) - \sin\phi\}, \tag{2.6}$$

$$x(t) = \int_0^t v(t)\mathrm{d}t = \frac{E_0 e}{m_e\omega^2}\{\cos(\omega t + \phi) - \cos(\phi) + \sin(\phi)t\}. \tag{2.7}$$

The kinetic energy of the electron can be calculated from Eq. 2.6, which, after averaging over an optical cycle, results in the ponderomotive potential introduced before. As a consequence $U_p \propto I$. The electrons can either drift away or recollide with the parent ion, which can be found by calculating the trajectory of the electrons for different phases ϕ in Eq. 2.7. This is discussed in full for instance in [11].

2.1.3 Recombination

When an electron recollides with its parent ion, three different processes might occur with different probability: (i) elastic rescattering of the electron on the ion; (ii) inelastic scattering with excitation or ionization of bound electrons in the ion; (iii) recombination of the electron with the parent ion. If recombination occurs a photon is emitted (Fig. 2.2c), having the combined kinetic energy of the electron and the ionization potential of the ion

$$\hbar\omega = I_p + E_{\text{kin}}. \tag{2.8}$$

According to Eq. 2.6 the kinetic energy of the electron depends on the time it spends in the laser field. This obviously depends on the phase ϕ of the laser field at the time when the atom was ionized. Solving the equation of motion of the electron (Eq. 2.6) with $x(t) = 0$ for different phases ϕ one gets a maximum kinetic energy for $\phi \approx 18°$. Thus $E_{\text{kin}} = 3.17 \cdot U_p$ is the upper limit for the energy of the emitted photon by the electron. Together with the ionization potential of the atom one gets the so-called cut-off energy

$$\hbar\omega_{\text{cutoff}} = I_p + 3.17U_p. \tag{2.9}$$

Despite the simplicity of the three-step-model and its deterministic connection of classic and quantum mechanical effects, it is able to describe the features observed in HHG. The structure of the cut-off, the plateau behavior (Fig. 2.1), and the existence of only odd harmonics, due to every half-cycle causing recollisions, are well explained. A more rigorous quantum mechanical model for HHG was developed by Lewenstein et al. [12], which will not be discussed in depth in this thesis.

Other general properties one has to mention when dealing with HHG is that the light produced is highly coherent and emitted in a beam having typically a smaller divergence compared to the driving laser beam [13, 14]. The photon flux generated with HHG is low compared to synchrotrons or free-electron lasers. Limiting for the photon flux is the low conversion efficiency. The highest reported HHG conversion efficiencies are in range of $\approx 10^{-5}$ [15–17]. As with other nonlinear optical processes phase matching, i.e. matching the phase velocity of the fundamental light with the phase velocity of the generated harmonics, is one of the main limiting issues. Another important issue for microscopy with high harmonics beside the flux is the bandwidth of each harmonic. The bandwidth is getting narrower the more optical cycles contribute to the signal. Thus long driving pulses generate narrowband harmonics as they are essential for high resolution imaging as will be shown in Sect. 2.2.4. The pulse duration of the harmonics emitted is intrinsically shorter than the driving laser pulse and can reach the attosecond level [18].

2.2 Coherent Diffraction Imaging Theory

After the introduction of HHG for frequency up-conversion, the topic of this section will lead towards the application of those unique light pulses for microscopy. First, the basic principles of diffraction of light by matter will be discussed. Then specific problems and methods used in CDI to solve the so-called *phase problem* will be introduced. At the end of this section a closer look at geometric considerations one has to bear with when targeting the implementation of a CDI experiment in the lab will be taken. It is worth mentioning that direct imaging with Fresnel zone plates [19] has been done for decades at synchrotrons [20] and more recently using HHG sources [21], however, the resolution is limited due to the limited resolution in electron beam lithography for producing the zone plates. Further, the bandwidth of typical HHG pulses is limiting the usability of Fresnel zone plates.

2.2.1 Coherence Properties of a Light Field

An important measure for imaging purposes is the coherence. Coherence is a property of the light field giving information about how well defined the phase relation between different parts of the electromagnetic radiation is either in space or time. These are referred to as the spatial and temporal coherence, respectively. Spatial coherence describes the phase relation between two separated points in a plane perpendicular to the propagation direction [22]. Temporal coherence on the other hand describes a defined phase relation between two points separated in time, i.e. parallel to the propagation direction. At this stage one can already conclude that plane waves have perfect spatial coherence, since they have the same phase at all points in planes perpendicular to the propagation direction. In reality however, one always has a source of light that has a finite size, i.e. it emits spherical waves which can be approximated as plane waves only for large distances from the source. Thus, as shown in the seminal book from Attwood [22], one can relate the spatial coherence of a light source to its spatial size and its emission characteristics, i.e. the divergence $\Delta\theta$ of the radiation emitted. A measure for this is the transverse coherence length ξ_t

$$\xi_t = z\Delta\theta = \frac{z\lambda}{2\pi d}, \tag{2.10}$$

where z is the distance from the source to the plane of observation, λ is the wavelength and d is the source size. A way to measure spatial coherence is to measure with a

Young's double slit with a variable distance between the slits. The wider the slits are separated, the wider is the separation of the sampling points on the wave that impinges on the double slit. If the separation between the slits becomes greater than ξ_t one observes decreasing contrast on the fringes in the interference pattern in the far-field behind the double slit. In a similar manner, for coherent imaging one finds that spatial coherence is limiting the size of typical samples that can be imaged. This is because if samples are significantly larger than ξ_t one cannot expect clearly resolved fringes in the far-field and, thus, cannot resolve the phase as will be presented in Sect. 2.2.3. What makes the estimation of the spatial coherence according to Eq. 2.10 difficult in the practical application of CDI is the fact that typically the light from a source is refocused and the sample is placed in the focus. Hence z would be zero and one cannot use Eq. 2.10, which only applies for a divergent light field. Thus a more useful interpretation, but less measurable, is that the spatial coherence is related to the beam quality, i.e. one should optimize for a Gaussian-shaped focal spot with an ideally flat phase front.

Temporal coherence on the other hand is related to the bandwidth of the light field. This can be intuitively understood again by analyzing a Young's double slit. Since different wavelengths are diffracted under different angles one gets more smearing (broadening) of the fringes and therefore less contrast the larger the bandwidth of the incoming light field becomes. The effect increases for larger diffraction angles, i.e. larger distance from the center of the pattern. The temporal coherence can be quantified by the longitudinal coherence length ξ_l

$$\xi_l = \frac{\lambda^2}{2\Delta\lambda}. \tag{2.11}$$

$\Delta\lambda$ denotes the full width half maximum (FWHM) bandwidth of the source.

Often temporal coherence measurements are discussed with Michelson interferometers in the way that changing the path length of one interferometer arm more than ξ_l the interference is lost due to the missing phase relation between the interfering light fields. Here temporal coherence is discussed at the far-field interference pattern of a double slit. Comparing this to the discussion on spatial coherence one finds that temporal coherence effects the far-field pattern of the object, i.e. the double slit, in a similar way as spatial coherence. The only difference is that temporal coherence is more limiting at higher diffraction angles, i.e. the achievable resolution, while spatial coherence is limiting the field of view. Thus in a CDI experiment both coherence measures are of special importance.

2.2.2 Diffraction of Light Waves at Matter

In this section the basic concepts of diffraction of light waves at matter will be discussed. This field is very complex and cannot be covered in depth within this thesis. Hence, in this work the focus will be on a few principles and a few important

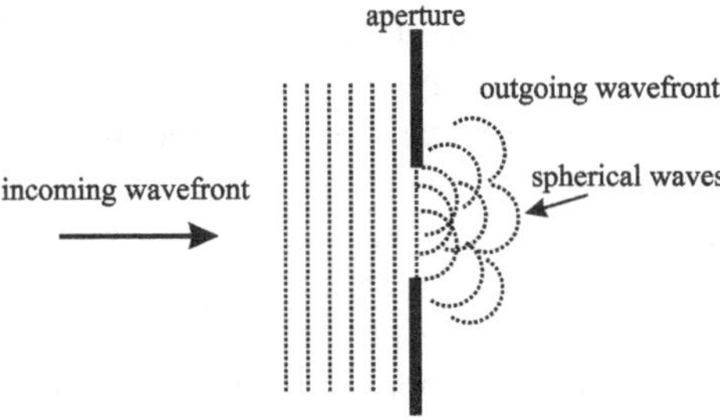

Fig. 2.3 The Huygens-Fresnel principle describes the diffraction of light as the superposition of spherical waves that emerge from an object. In the presented case this is a simple aperture

formulas to illustrate the main effects that are relevant for CDI. A more detailed discussion and further reading can be found for instance in the textbook of Jackson et al. [23]. Here the notation given in [24] will be followed, hence electric fields will be denoted by U.

An intuitive picture of diffraction is given by the Huygens-Fresnel principle (Fig. 2.3). It states that every part of a wavefront emerging from an obstacle can be considered as the center of a spherical wave. Thus the field at later points, after a certain propagation, can be described by the superposition of all spherical waves that propagated the given distance. The superposition of coherent wavefronts usually gives rise to interference effects and hence produces a certain intensity distribution that is usually called a *diffraction pattern*. So the Huygens-Fresnel principle describes the principal effects, however, the drawback is obvious, since one would need an infinite number of waves to be propagated and interfered to determine a diffraction pattern of an actual object. Another consequence of the Huygens-Fresnel principle is that a diffraction pattern that one can expect behind an aperture depends on the distance between the aperture and the plane of observation. For distances close to the aperture ($z \sim a$, where a is the aperture size), geometrical optics can be applied to describe the field structure. Going further away the diffraction pattern evolves quickly as a function of the distance since the interference of the spherical waves starts to build up. This is called Fresnel diffraction. For distances far from the object the spherical waves can be estimated as plane waves having different phases, thus the diffraction pattern does not change but merely increases in size. This case is called Fraunhofer diffraction. The transition from Fresnel to Fraunhofer diffraction is not abrupt. One can estimate the distance z for which the Fresnel regime changes into the Fraunhofer regime as

$$z \approx \frac{a^2}{\lambda}, \tag{2.12}$$

where a is the aperture size and λ is the wavelength.

This is a rather qualitative view on diffraction. In next part of this section a quantitative description of the diffraction of an arbitrary object will be derived. As shown in [24] one can derive

$$U(P) = \frac{1}{4\pi} \iint\limits_{S} \left[U \frac{\partial}{\partial \mathbf{n}} \left(\frac{\exp(-\mathbf{ikr})}{r} \right) - \frac{\exp(-\mathbf{ikr})}{r} \frac{\partial U}{\partial \mathbf{n}} \right] \mathrm{d}S \tag{2.13}$$

from the Huygens-Fresnel principle, where U is a solution of the Helmholtz equation at a point of observation P, with $\mathbf{n}$ being a unit vector normal to the emitting surface S, $\mathbf{r}$ being the vector from S to P and $\mathbf{k}$ being the wavevector. Equation 2.13 is called the Kirchhoff diffraction integral. If one now assumes to have an aperture on an otherwise opaque screen that is illuminated from one side and the observation plane is in the halfspace on the other side, one can derive [24] the Fresnel-Kirchhoff diffraction formula

$$U_K(P) = \frac{\mathrm{i}U_0}{\lambda} \iint\limits_{\Sigma} \frac{\exp[-\mathbf{ik}(\mathbf{r}+\mathbf{s})]}{\mathbf{rs}} \frac{[\cos(\mathbf{n},\mathbf{s}) - \cos(\mathbf{n},\mathbf{r})]}{2} \mathrm{d}S. \tag{2.14}$$

In Eq. 2.14 $\mathbf{s}$ denotes the vector from the aperture $\sum$ to the source and $\mathbf{r}$ denotes the vector from the aperture to the point of observation P. The integration is done over $\sum$ within the halfspace S enclosing P. The Fresnel-Kirchhoff diffraction formula is still somewhat too general [24] and contains some mathematical irregularities. RAYLEIGH and SOMMERFELD used certain symmetries and assumptions to simplify Eq. 2.14 further. Using their simplifications and introducing a Cartesian coordinate system (Fig. 2.4) one can now calculate the light field U_2 at a point P_2 in the observation plane

$$U_2(x_2, y_2) = \frac{\mathrm{i}}{\lambda} \iint\limits_{-\infty}^{\infty} U_1(x_1, y_1) \frac{\exp(-\mathbf{ikr})}{r} \cos(\mathbf{n}, \mathbf{r}) \mathrm{d}x_1 \mathrm{d}y_1, \tag{2.15}$$

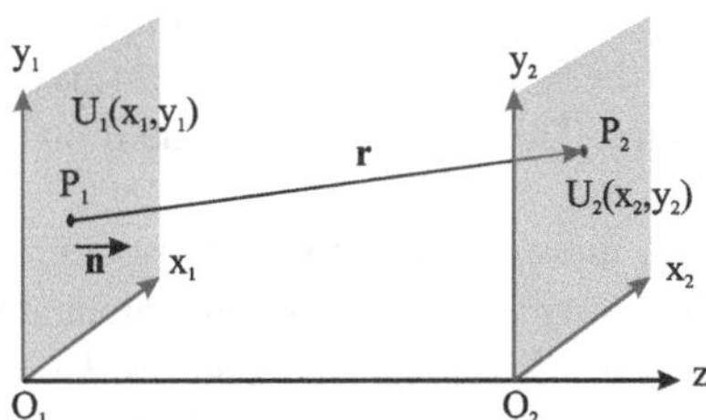

Fig. 2.4 Definition of the coordinate system. U_1 is the light field at point P_1 in the diffraction plane (x_1, y_1) and U_2 is the light field in the observation plane (x_2, y_2) at point P_2

where U_1 is the light field in the diffraction plane and $\cos(\mathbf{n}, \mathbf{r})$ is the cosine of the angle between the normal to the diffraction plane $\mathbf{n}$ and the direction of observation $\mathbf{r}$.

The factor $\frac{\exp(-i\mathbf{kr})}{r}$ describes a spherical wave originating at point P_1 and demonstrates the Huygens-Fresnel principle. Provided one knows the light field within an aperture U_1 one can calculate the diffraction pattern anywhere behind the aperture if $r \gg \lambda$ which is one of the main assumptions when deriving Eq. 2.15. For typical CDI experiments one expects a plane wave impinging on the target, thus U_1 is well defined. But also in the case of Fresnel CDI [25] one can still use Eq. 2.15 by assuming a spherical wavefront illuminating an aperture. Using the coordinate system (Fig. 2.4) the distance r between P_1 and P_2 can be expressed in Cartesian coordinates

$$r^2 = z^2 + (x_2 - x_1)^2 + (y_2 - y_1)^2 = z^2 \left[1 + \frac{(x_2 - x_1)^2 + (y_2 - y_1)^2}{z^2} \right]. \quad (2.16)$$

If the observer resides near the optical axis, which is the typical situation in CDI, one can use $(x_2 - x_1)^2 + (y_2 - y_1)^2 \ll z^2$ to simplify Eq. 2.16 to

$$r \approx z \left[1 + \frac{(x_2 - x_1)^2 + (y_2 - y_1)^2}{2z^2} \right]. \quad (2.17)$$

This paraxial approximation is called the Fresnel approximation. In this case the cosine in Eq. 2.15 becomes unity and using Eq. 2.17 one can reduce Eq. 2.15 to

$$U_2(x_2, y_2) = \frac{\mathrm{i}\exp(-\mathrm{i}kz)}{\lambda z} \iint_{-\infty}^{\infty} U_1(x_1, y_1) \exp\left[-\mathrm{i}k \frac{(x_2 - x_1)^2 + (y_2 - y_1)^2}{2z} \right] \mathrm{d}x_1 \mathrm{d}y_1. \quad (2.18)$$

This formula yields the diffraction pattern in the paraxial approximation, which is also called the Fresnel diffraction pattern or the near-field. Inspecting Eq. 2.18 reveals that the phase in the observation plane has a quadratic dependence on the position. This is the reason as to why the Fresnel diffraction regime is complicated and difficult to handle in imaging systems. This is particularly true if one deals numerically with such geometries.

As mentioned earlier, a more orderly diffraction pattern is expected when the observation plane is further away from the diffraction plane. In this case the quadratic terms of the coordinates in the diffraction plane in Eq. 2.17 can be neglected since $z^2 \gg x_1^2, y_1^2$. One only keeps the mixing terms and gets

$$r \approx z \left[1 + \frac{x_2^2 + y_2^2}{2z^2} - \frac{x_1 x_2 + y_1 y_2}{z^2} \right]. \quad (2.19)$$

Using this so-called Fraunhofer approximation and plugging it into the Rayleigh-Sommerfeld diffraction formula (Eq. 2.15) one gets

$$U_2(x_2, y_2) = \frac{\mathrm{i}\exp(-\mathrm{i}kz)}{\lambda z}\exp\left(-\mathrm{i}k\frac{x_2^2+y_2^2}{2z}\right)$$
$$\iint_{-\infty}^{\infty} U_1(x_1, y_1)\exp\left[\frac{\mathrm{i}k}{z}(x_1x_2+y_1y_2)\right]\mathrm{d}x_1\mathrm{d}y_1. \quad (2.20)$$

Inspecting this Fraunhofer diffraction formula one finds that a nonlinear phase variation in the observation plane is not arising. Instead one gets a steady pattern that just expands in size over the propagation distance z. The integral essentially denotes the Fourier transform of the source field U_1.[2] This is a very important result for CDI, since using numerical implementations of the Fourier transform, e.g. Fast Fourier Transforms (FFT), one can numerically switch between the diffraction pattern in the far-field and the source field at an aperture. This is the basis for iteratively solving the phase problem as will be explained in the subsequent section.

In summary the discussion started with the general Huygens-Fresnel principle that essentially describes diffraction as the sum of point sources originating at the surface of the diffracting object. Putting this into a mathematical description by integrating over all spherical waves that leave the object, a solution that obeys the Helmholtz equation for wave propagation was found. By restricting to one halfspace and using paraxial and a few further assumptions the Fresnel diffraction formula was derived. The Fresnel diffraction formula (Eq. 2.18) is valid at any distance significantly larger than the wavelength behind an illuminated aperture a. However, for even larger distances $z \gg \frac{a^2}{\lambda}$ the Fraunhofer diffraction formula (Eq. 2.20) was found. It relates the far-field and the source field by the Fourier transform. Thus the coordinates in the far-field can be related to spatial frequencies q by

$$q_x = x/(z\lambda) \quad \text{and} \quad q_y = y/(z\lambda). \quad (2.21)$$

In terms of interpreting diffraction as an elastic scattering of photons one can interpret these spatial frequencies as the components of the momentum transfer vector $\mathbf{q}(q_x, q_y, q_z)$. The photons exiting the diffraction plane $\mathbf{k}_{\text{out}}$, which will later be denoted as the object plane, experience a higher momentum change the more the propagation directions differ from the incidence direction of the photons defined by the wavevector $\mathbf{k}_{\text{in}}$ of the incoming wave. Hence

$$\mathbf{q} = \mathbf{k}_{\text{out}} - \mathbf{k}_{\text{in}} \quad (2.22)$$

[2] In literature this field is also often referred to as the *exit surface wave*.

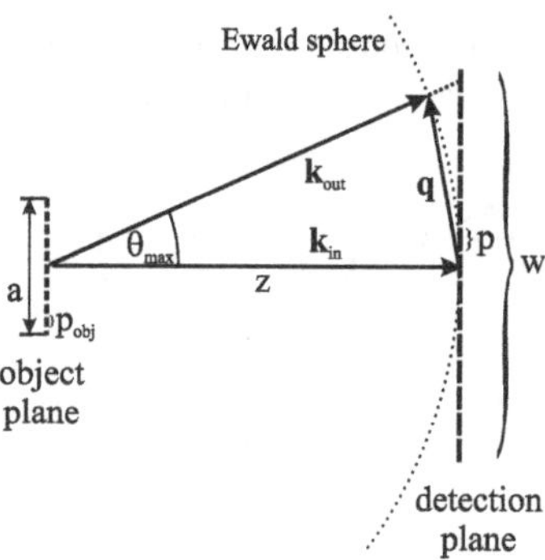

Fig. 2.5 Scattering of an incident coherent beam $\mathbf{k}_{in}$ on a planar object in the object plane (diffraction plane). The wavevector of the exiting photons $\mathbf{k}_{out}$ ends on the Ewald sphere. Since one measures a diffraction pattern in the detection plane, one can interpret the spatial frequencies as the projection of the momentum transfer vector $\mathbf{q}$ on the corresponding axis. The object plane is characterized by its spatial extent a and a sampling size, i.e. the object pixel size, p_{obj}. The detection plane is likewise characterized by the detector size w and the pixel size p. Please note that one can already see in this sketch that the Ewald sphere is projected onto the planar detector. For high scattering angles, e. g. denoted by θ_{max}, as they occur for high numerical aperture measurements, a correction of this effect is necessary. Details can be found in Sect. 3.4

gives the resulting momentum transfer vector.[3] Energy conservation imposes $|\mathbf{k}_{out}| = |\mathbf{k}_{in}|$. Hence, all scattering vectors, and thus the diffraction pattern, lie on a sphere, which in crystallography is called the Ewald sphere [27]. The Ewald sphere has thus the radius $1/\lambda$, since $k \propto 1/\lambda$. The scattering geometry is depicted in Fig. 2.5. Throughout this thesis it will be assumed that the Ewald sphere has a radius $1/\lambda$.[4] The momentum transfer $q = |\mathbf{q}|$ for a given scattering angle θ thus becomes [28][5]

$$q = \frac{2}{\lambda} \sin\left(\frac{\theta}{2}\right). \tag{2.23}$$

2.2.3 *The Phase Problem*

In the preceding section it was found that the object plane, i.e. the source field, and the detection plane are Fourier conjugates in the Fraunhofer approximation, which

[3] One can compare Eq. 2.22 with the well-known momentum transfer equation from elastic scattering $\Delta\mathbf{p} = \mathbf{p} - \mathbf{p}_0$ by multiplying both sides in Eq. 2.22 with $\hbar = h/2\pi$ where h is Planck's constant. If one now uses the de Broglie relation $\mathbf{p}_0 = \hbar\mathbf{k}_{in}$ and $\mathbf{p} = \hbar\mathbf{k}_{out}$ for the incident and scattered particle, respectively, it becomes clear that $\hbar\mathbf{q} = \hbar\mathbf{k}_{out} - \hbar\mathbf{k}_{in}$ is the momentum transfer vector [26].

[4] For the sake of being consistent with *most* of the literature on CDI. Hence, $|\mathbf{k}| = 1/\lambda$ is used.

[5] This equation comes from an isosceles triangle, where $c = 2a\sin(\gamma/2)$. Comparing it to the well-known Bragg's law one finds that Eq. 2.23 is equivalent to a volume grating being tilted about half the diffraction angle θ.

is almost always fulfilled for soft X-ray imaging due to the short wavelength. Thus one could retrieve the object except for a constant phase factor by a simple Fourier transform if one could sample the amplitude and the phase in the far-field. However, in all experiments that will be presented throughout this thesis an XUV sensitive charge coupled device detector (CCD, details in Chap. 3) is used to measure the diffraction patterns. Unfortunately, there are no detectors that can measure phase and amplitude spatially resolved simultaneously; instead a CCD measures only intensities, i.e. $I \propto UU^* = |U|^2$. Hence the phase information is typically lost in CDI experiments, a predicament commonly called the *phase problem*.

In 1952 Sayre published a seminal article [29] where he states that in principle the phase information can be recovered from the diffraction intensities if they are sampled densely enough. The idea behind this is the Shannon theorem [30], which essentially states that one can retrieve the phase of a signal if it is sampled at twice its frequency, the so-called *Nyquist frequency*. As shown before, one can think of the plane of a diffraction pattern as a plane of spatial frequencies. Thus the implication from Sayre's work that he continued years later [31] is that one can retrieve the phase if the continuous X-ray diffraction pattern caused by an non-periodic object is sampled sufficiently dense. This is called the oversampling of the diffraction pattern. Bates discussed in [32] that the solution of this inversion problem would be almost always unique. If one considers a sample with a diameter a one can determine the spatial Nyquist frequency to $f_{\text{Nyquist}} = a/(z\lambda)$. This would be the spatial frequency with which one would have to sample the diffraction pattern with, in order to achieve a direct conjugate relation between object plane and the diffraction pattern. If the phase problem is considered as a set of equations one would have N equations in both planes, i.e. N pixels considering a CCD. However, since every detection spot is characterized by an amplitude and a phase one has $2N$ unknowns and due to the phase problem only N, i.e. the amplitudes $\sqrt{I}$, knowns. To solve this system of equations one can, however, sample the diffraction pattern with $2f_{\text{Nyquist}}$, i.e. having two pixels per spatial Nyquist frequency in terms of a CCD, and have $2N$ knowns ($2N$ amplitudes). If one now uses that there are $2N$ unknowns (N amplitudes and N phases) in the object plane, one can in principle retrieve the phase [33]. In terms of an FFT this means that the object plane has a certain zero padding around the object, where amplitude and phase are known to be zero. This is depicted as two-dimensional case in Fig. 2.6.

To quantify the oversampling of a diffraction pattern one can introduce an oversampling ratio σ, which one can define in the object plane as

$$\sigma = \frac{\text{region with electron density} + \text{region without electron density}}{\text{region with electron density}}. \tag{2.24}$$

The formulation *electron density* comes from X-ray diffraction on atoms, where the dipoles of the electrons in an atom are the sources of the spherical wavelets [23]. To be consistent with most CDI literature the electron density ρ will further be used as a substitute for the source terms U in the previous section. So even for an aperture,

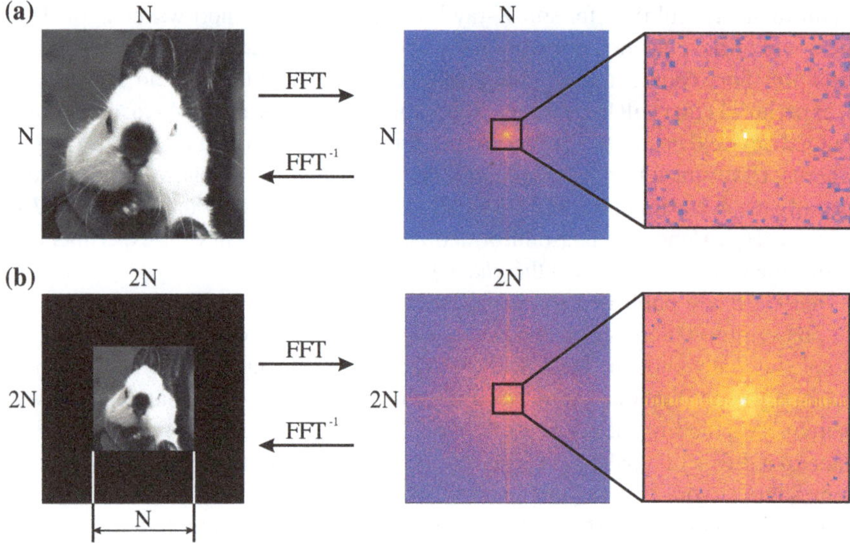

Fig. 2.6 Oversampling as condition for solving the phase problem. **a** An object that fills N by N pixels in the object plane results in a N by N pixels diffraction pattern in the Fourier transform plane. **b** If one, however, adds a zero-density region around the object, i.e. zero-padding, to $2N$ by $2N$ pixels one gets the corresponding $2N$ by $2N$ diffraction pattern. This represents a twofold oversampled diffraction pattern with $2N$ by $2N$ known amplitudes, which allow to solve the phase problem (the amplitude in the Fourier transform plane is plotted on the *right panels* on a logarithmic scale). This is because there are only N by N complex-valued unknowns in the object plane and all other pixels in the zero-density region can be set to zero. For convenience the arrays are shown at the same size, in reality the arrays in (**b**) are four times larger

where the field in the object plane can be reconstructed, one can recover the object itself through Babinet's princple [24].

According to Sayre $\sigma \geq 2$ is necessary to solve the phase problem. In the detection plane this corresponds to the number of pixels used to sample a speckle in the diffraction pattern. In two dimensions $O = \sqrt{\sigma}$ the linear oversampling degree in the detection plane can be introduced [34]

$$O = \frac{\lambda z}{ap}, \tag{2.25}$$

with λ being the wavelength, p the width of a detector pixel, z the distance between object and detection plane and a the largest spatial extent of the object. According to Eq. 2.21 this corresponds to the highest spatial frequency observed from an object related to the size of a detector pixel.

In this section it was shown that the phase problem can in principle be solved, if sufficient oversampling in the diffraction pattern is maintained. The ambiguity that is left and cannot be recovered is a lateral translation of the object, a complex conjugation, a spatial inversion and an absolute phase shift [32]. One should note that all these considerations are only valid for the Fraunhofer regime, because the nonlinear phase evolution in the Fresnel regime, as pointed out in Sect. 2.2, would not satisfy a linear sampling theorem [34]. To maintain sufficient oversampling in an experiment, careful considerations regarding the experimental geometry are necessary. This will be covered in the following section.

2.2.4 Geometric Considerations

In the preceding section it was found that oversampling (Eq. 2.25) is a crucial requirement for CDI experiments in order to retrieve the object plane from a far-field diffraction pattern. The far-field condition was given in Eq. 2.12. In Sect. 2.2.1 the coherence properties of a light source were investigated. Hence it is beneficial to bring the findings together and have a look at the restrictions to the experimental geometry and the properties of the light source that arise from these findings. Spence et al. present a comprehensive discussion [35] on the coherence requirements of a source for proper CDI. For the transverse coherence length they found that

$$\xi_t > 2a, \tag{2.26}$$

where ξ_t is the transverse coherence length as discussed in Sect. 2.2.1 and a is the spatial extent of the object. The reason for this is that the autocorrelation function Γ of an object o

$$\Gamma(\mathbf{r}) = o(\mathbf{r}) \otimes o^*(-\mathbf{r}), \tag{2.27}$$

where $\otimes$ denotes the convolution operator, has twice the size of its spatial extent a [28]. That means that according to [35] the transverse coherence must at least be maintained over the size of the autocorrelation Γ, because the unknown phases are compensated by collecting coherent diffraction data from a plane at least twice as big as the sample. For higher oversampling O one can thus generalize Eq. 2.26 to

$$\xi_t > Oa. \tag{2.28}$$

For the temporal (longitudinal) coherence similar estimations can be made, i.e. two light rays originating from opposite sites of the sample should still interfere on the detector. Miao et al. empirically found [34] a relation between the bandwidth $\Delta\lambda$ of a light beam, the achievable resolution Δr and the oversampling O:

$$\frac{\lambda}{\Delta\lambda} \geq \frac{Oa}{\Delta r} \Longrightarrow \Delta r \geq \frac{Oa\Delta\lambda}{\lambda}. \tag{2.29}$$

Hence the resolution is limited by the bandwidth,[6] i.e. the fringes smear out as was qualitatively already discussed in Sect. 2.2.1. At the same time one sees from Eqs. 2.28 and 2.29 that for a higher oversampling O one needs a higher temporal coherence, i.e. a smaller bandwidth $\Delta\lambda$, and a higher transverse coherence ξ_t. Since no additional resolution in the reconstruction is gained from higher oversampling, it can be concluded from this result to aim for $O \gtrsim 2$. If not avoidable, one can account for partial coherence in the phase retrieval algorithm (Sect. 2.3) [36].

If one now considers a diffraction experiment where the diffraction pattern is sampled on a regular N_x by N_y grid, one can assign every sampling point, i.e. each pixel, a corresponding sampling interval Δq_x and Δq_y. The corresponding sampling interval of each single pixel at θ can be calculated using Eq. 2.23. From this one can deduce the field of view (FOV) in each axis to [28]

$$L_x = \frac{1}{\Delta q_x}, \qquad L_y = \frac{1}{\Delta q_y}. \tag{2.30}$$

At the same time the highest measured momentum transfer on the q_x-axis $q_{x,\max}$ results in the smallest resolvable period Δr_x in the corresponding object plane axis[7]

$$\Delta r_x = \frac{1}{2q_{x,\max}}. \tag{2.31}$$

The same is of course valid for the q_y/y-axis and typically has the same result in a transmission CDI experiment where the detector is centered behind the sample. Hence, for a given detector width w and a detector sample distance z (see Fig. 2.5) one can determine the highest recorded scattering angles $\theta_{\max} = \tan^{-1}[w/(2z)]$ and use these to calculate the highest detectable momentum transfer as[8]

$$q_{\max} = \frac{2}{\lambda}\sin\left(\frac{\theta_{\max}}{2}\right) = \frac{2}{\lambda}\sin\left[\frac{1}{2}\tan^{-1}\left(\frac{w}{2z}\right)\right]. \tag{2.32}$$

[6] Provided that diffraction data is measured to sufficiently high momentum transfers, such that in principle a higher resolution could be obtained considering perfect coherence properties. See the following pages for the fundamental resolution limit.

[7] It is assumed that the zero deflection point, i.e. $|q| = 0$, resides in the center of the detector. The factor 2 comes from the fact that the pattern is sampled from $-[(N_x - 1)/2]\Delta q_x$ to $[N_x/2]\Delta q_x$ on the detector and that the FFT conserves the total amount of pixels, i.e. the object plane is also sampled by N_x times N_y pixels.

[8] This refers to the midpoints from the diagonal edges of the CCD. Sometimes in literature the highest momentum transfer in the edges of the detector is mentioned. It is thus a factor of $\sqrt{2}$ larger, but this yields no additional resolution because the highest momentum transfer with respect to the x- and y-axis is still the same.

Thus the smallest resolvable period becomes

$$\Delta r = \frac{\lambda}{4\sin\left[\frac{1}{2}\tan^{-1}(\frac{w}{2z})\right]}. \tag{2.33}$$

For small angles θ_{max} this compares to the well-known Abbe limit (Eq. 1.1), if $\mathrm{NA} = \sin[\tan^{-1}(\frac{w}{2z})]$ is considered as the numerical aperture of the system. Using the small angle approximation $\sin[\tan^{-1}(w/[2z])] \approx w/(2z)$ yields

$$\Delta r = \frac{z\lambda}{w} = \frac{z\lambda}{pN} = p_{\mathrm{obj}} \tag{2.34}$$

if a CCD having N pixels with a pixel width p is considered. Due to the principle of conservation of the number of pixels one can also interpret this as the pixel size p_{obj} in the object plane. This is also referred to as the *half-pitch distance* and it gives an estimate of the smallest details in an object that can be resolved provided the diffraction pattern can be measured with a sufficient signal to noise ratio at the edge of the detector.

2.3 Iterative Phase Retrieval Methods

In the previous sections diffraction of light waves on a sample was investigated and the phase problem (Sect. 2.2.3) was found, which needs to be solved in order to reconstruct the object. As pointed out by Sayre in 1952 [29], one can in principle solve the phase problem if oversampled diffraction patterns are measured. However, it took almost 20 years until Gerchberg and Saxton came up with a first algorithm [37] for the phase retrieval in a diffraction pattern, at this time for electron microscopes. In 1978 Fienup then introduced an improved version [38] of the Gerchberg-Saxton algorithm, which is known as the error-reduction algorithm. He introduced a support constraint, i.e. setting an area outside of the expected object to zero, and showed that this results in a more stable phase retrieval. Moreover, he introduced a positivity constraint, since a physical object's complex electron density cannot have a negative amplitude. The error-reduction algorithm, however, suffered from stagnation in local minima during the reconstruction and thus needed further improvement, which resulted in the hybrid input-output algorithm (HIO) [38]. In the HIO algorithm the amplitude that is projected outside of the support is used as an error, which one tries to reduce from iteration to iteration. The principal scheme for an iterative phase retrieval is depicted in Fig. 2.7. For the initialization of the algorithm (Fig. 2.7a) the measured amplitudes, i.e. $\sqrt{I(\mathbf{q})}$, and random phases are used. An initial support function S can be determined from the diffraction pattern if not otherwise known. This issue will be tackled later in this section.

The goal of the phase retrieval is to reconstruct the complex wave function $U(\mathbf{r})$, as discussed in Sect. 2.2.2, that is exiting the object plane. This is equivalent to

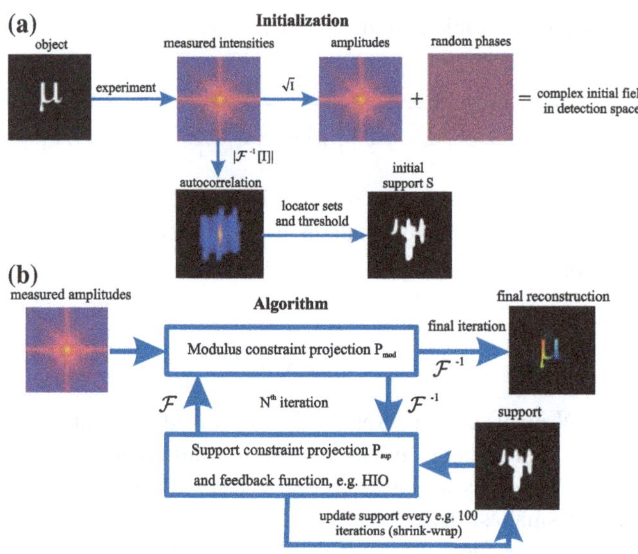

Fig. 2.7 Principal scheme for an iterative phase retrieval. **a** In a diffraction experiment one typically measures the intensities of the diffraction pattern in the far-field, i.e. the Fourier transform of the object. Since nothing is known about the phases one typically seeds the algorithm with random phases. From the modulus of the inverse Fourier transform of the measured intensities one obtains a function proportional to the autocorrelation, which can be used to obtain an initial support S that encloses the object. So-called locator sets can be used to enhance this first support estimate [39]. **b** After having seeded the algorithm as explained in (**a**) one iterates between detection plane and object plane using corresponding Fourier transforms. In the detection plane the modulus constraint is enforced, i.e. taking the measured amplitudes and keeping the phases. Likewise, in the object plane one enforces the support constraint by using a proper feedback function, which could for instance be the HIO formulation. The shrink-wrap method [40] is used to enhance the support every e.g. 100 iterations. The final reconstruction is depicted as complex-valued plane, where the hue and brightness encode the phase and amplitude, respectively. The linear phase ramp, visualized by the color transition in the final reconstruction, comes from a non-perfect centered diffraction pattern

reconstructing the object if one considers a coherent plane wave impinging on the object. In the detector plane one has the complex wave $\tilde{U}(\mathbf{q}) \propto \mathcal{F}[U(\mathbf{r})]$, from which one can only measure the intensity $I(\mathbf{q}) \propto \tilde{U}(\mathbf{q})\tilde{U}^*(\mathbf{q})$. Thus the modulus constraint (Fig. 2.7b) in the n-th iteration of the algorithm in the detector plane is defined as

$$\tilde{U}_{n+1}(\mathbf{q}) = \sqrt{I(\mathbf{q})}\,\exp[\mathrm{i}\,\arg\{\tilde{U}_n(\mathbf{q})\}], \tag{2.35}$$

i.e. one keeps the phases and replaces the amplitude by the square root of the measured intensities. Now one can think of the back transformation to the object plane as a projection, i.e. the modulus constraint projection, which is denoted as P_{mod}. This allows to write the modified wave function in the object plane in terms of the projection as

$$U_{n+1}(\mathbf{r}) = P_{\mathrm{mod}}U_n(\mathbf{r}) = \mathcal{F}^{-1}[\tilde{U}_{n+1}(\mathbf{q})]. \tag{2.36}$$

Next a projection in the object plane is applied, which is called the support projection P_{sup}. The support S is the region inside the object, i.e. surrounded by a zero-density region. One can write this projection as

$$P_{\text{sup}} U_{n+1}(\mathbf{r}) = \begin{cases} U_{n+1}(\mathbf{r}) & \in S \\ 0 & \notin S \end{cases}. \tag{2.37}$$

The beauty of these projections is that one can use them to write down a certain iterative algorithm in one formula. For instance the error-reduction algorithm can be written as

$$U_{n+1}(\mathbf{r}) = P_{\text{sup}} P_{\text{mod}} U_n(\mathbf{r}) \tag{2.38}$$

or the HIO as

$$U_{n+1}(\mathbf{r}) = \begin{cases} P_{\text{mod}} U_n(\mathbf{r}) & \in S \\ U_n(\mathbf{r}) - \beta P_{\text{mod}} U_n(\mathbf{r}) & \notin S \end{cases}, \tag{2.39}$$

where β is a feedback parameter with typical values between 0.1 and 1. Furthermore, sometimes additional constraints are enforced on the object plane. For example one can constrain that the object must be real-valued and positive $U_{n+1} \in \mathbb{R}^+$ in case of a non-absorbing aperture. In case of a pure phase-object one can use $U_{n+1} = \exp[\mathrm{i} \arg\{U_{n+1}\}]$ as an additional constraint. Even though this requires some knowledge about the object, it significantly decreases, if available, the search space for the algorithm and consequently speeds up the convergence. Over the years many different modifications to the HIO were introduced with each following a different goal. A comprehensive overview and comparison can be found in [40]. Mostly a mixture of algorithms is used within a reconstruction run, e.g. 100 iterations of HIO followed by a few error-reduction steps [28].

A realization of a noise-robust HIO dealing with noise[9] in the recorded diffraction pattern is presented in [41]. This realization was used for reconstructions in this thesis apart from the unmodified HIO. Another realization that was implemented and worked well on experimental data was Lukes relaxed averaged alternating reflection algorithm (RAAR) [42] which can be written as

$$U_{n+1}(\mathbf{r}) = \beta U_n(\mathbf{r}) + 2\beta P_{\text{sup}} P_{\text{mod}} U_n(\mathbf{r}) + (1 - 2\beta) P_{\text{mod}} U_n(\mathbf{r}) - \beta P_{\text{sup}} U_n(\mathbf{r}). \tag{2.40}$$

For all phase retrieval algorithms a tightly fitting support S is crucial. The support, however, is unknown in the general sense in microscopy. To get a first estimate of the support one can use that the autocorrelation of the object $o(\mathbf{r})$ is proportional to the inverse Fourier transform of the intensity diffraction pattern [39]

$$\Gamma(\mathbf{r}) = |\mathcal{F}^{-1}[I(\mathbf{q})]| \propto o(\mathbf{r}) \otimes o^*(-\mathbf{r}), \tag{2.41}$$

[9] Due to shot noise and electronic noise of the readout electronics.

which essentially means that one gets an outer bound for the object from the measured diffraction pattern. The autocorrelation can be practically thought of as translating the object $o(\mathbf{r})$ across itself in all directions and summing it up [39]. Hence the radius of the autocorrelation is always larger than the object. Using so-called *locator sets* [39], which basically use symmetries of the Fourier transform, one shifts $\Gamma(\mathbf{r})$ along a set of vectors and sums the resultant in a refined $\Gamma(\mathbf{r})$, see Fig. 2.7a. Proper thresholding of this refined $\Gamma(\mathbf{r})$ results in a support estimate S which is typically close to the actual object. Locator sets and the support determination from the autocorrelation will be used to reconstruct all objects in this thesis without any *a priori* knowledge.

Another important technique to improve the support during the iterations was introduced by Marchesini et al. and is known as the *shrink-wrap method* [43]. Shrink-wrap essentially updates the support after a given number of iterations based upon the current object plane reconstruction. In order to do this, one computes the modulus of the complex electron density in the object plane and applies a Gaussian filter and subsequently thresholds the result to determine a refined support. This method has proved to be very effective since it allows to get a tighter fit of the support to the actual object once the phases become more stable after a couple of iterations. It also helps to overcome the uncertainty of the first support estimation from the autocorrelation. Throughout this thesis shrink-wrap will be used for all reconstructions of experimental data. It is worth mentioning that an actual implementation of shrink-wrap introduces plenty of new parameters to the reconstruction, i.e. an amount of Gaussian blurring and the period of application. For most reconstructions in this thesis the Gaussian blurring is ramped down from e.g. 20 pixels standard deviation radius down to 1 and shrink-wrap is applied at about every 100th iteration.

When starting with random phases, one often finds deviating solutions in single HIO runs. Thus it is feasible to pick the best of the reconstructions after a certain number of iterations and average it with every other independent run and proceed further with the iterations. This technique is similar to genetic algorithms that are widely used for optimization problems [44]. In CDI this technique is termed *guided hybrid input-output* (GHIO) [45]. For some of the reconstructions in this thesis an implementation of GHIO was used. It worked well, especially if data is missing in the measured diffraction pattern, e.g. due to a beam stop that is used to block the bright central speckle.

The progress of the phase retrieval can be monitored by an appropriate error metric [46]. More important on the other hand is the success of the reconstruction in relation to the achieved resolution. As explained in Sect. 2.2.4 there is a fundamental resolution limitation set by the highest momentum transfer that can be measured. However, typically the magnitude of the signal in the Fourier domain decays quickly for increasing momentum transfer, thus in experiments the fringes measured at the edge of the CCD, if measured at all, are noisy. Hence, the resolution should be determined by the actual resolved diffraction pattern at the end of the phase retrieval, i.e. it should be checked what the highest momentum transfer that actually contributed to the reconstruction is. For this purpose Chapman et al. introduced the phase retrieval transfer function (PRTF) [28]

$$\mathrm{PRTF}(\mathbf{q}) = \frac{| < \tilde{U}_{\mathrm{final}}(\mathbf{q}) \exp(\mathrm{i}\phi_0) > |}{\sqrt{I(\mathbf{q})}}, \tag{2.42}$$

where $\tilde{U}_{\mathrm{final}}(\mathbf{q})$ is the final complex-valued reconstructed field in the detector plane without applying the modulus constraint in the last step. The angle brackets denote averaging over several independent reconstruction runs for which one needs a multiplicative phase constant ϕ_0 to adjust the phase offset between the runs to a common level. The PRTF is then typically integrated over shells of constant $|\mathbf{q}|$ and plotted over $|\mathbf{q}|$. This function drops to zero if there is no relation between the averaged reconstructed amplitudes and the measured amplitudes, hence the PRTF can be extrapolated to zero in the drop-off region to determine the highest $|\mathbf{q}|$ that contributed to the reconstruction.[10] Using the formulas given in Sect. 2.2.4 one can then determine the achieved resolution.

2.4 Digital In-line Holography

The phase problem (Sect. 2.2.3) is the main limitation to directly assess an object from its diffraction pattern. Moreover, CDI as presented in the previous section is limited to isolated objects. A way to overcome this limitation is the use of a well-defined reference wave that causes interference fringes within the diffraction pattern, which are related to the local phase at the detector. Thus, if sufficiently highly sampled, one can retrieve the phase indirectly. This principle is well-known as holography and was pioneered by Gabor in the later 1940s [47], winning him the Nobel prize in 1971. Holography nowadays is a broad field with plenty of applications and possible geometries. A comprehensive overview is given in the book from Toal [48]. The first holographic X-ray measurements were carried out using synchrotron radiation [49, 50]. Over the years this technique was improved for instance by using resonances in the material to achieve selective imaging [51] or ultrafast temporally resolved holography [52]. Moreover, this technique can be used to characterize the incoming X-ray field if one considers a known object [53, 54]. The first digital in-line holography experiments with table-top HHG sources were carried out using the divergent wavefront after a focus [13] and later using pinhole references [55]. If the object is placed in a divergent beam, e.g. behind a focus, one can also get a mixture between holography and CDI, which is termed *Fresnel CDI* [25, 56, 57]. Most of the X-ray holography experiments, however, use a separate source for the reference wave, which is typically a small pinhole [51, 58], an array of references [59, 60] or an extended reference [61–63]. This is somewhat limiting, because the reference needs to be placed at a well-defined distance from the object in order to acquire a hologram that can be directly inverted by a single Fourier transform, which is then termed *Fourier transform holography* [51, 62].

[10] Alternatively, one can threshold the PRTF at e.g. 0.5 or $1/e$ to determine the highest $|q|$.

In the general case of unknown, e.g. biological samples, in-line holography allows to assess the sample. In in-line holography the reference wave is the same as the wave that is scattered at the object, i.e. the illumination wave, while only a part of this wave is influenced by the object. For XUV wavelengths Gabor in-line holography was successfully used to write the hologram into PMMA for later analysis [64]. More practical in terms of microscopy is, however, digital in-line holography [65–67], where a detector captures the hologram and a computer carries out the reconstruction of an object image [68, 69]. From now on, the discussion in this thesis will be limited to digital in-line holography (DIH) only, since this type of holography was used for experiments in this work (Sect. 3.3).

The notation and discussion will follow an excellent overview given by Garcia-Sucerquia et al. [70]. In CDI one considers a plane wave impinging on the object causing a diffraction pattern. Since a plane wave in Gaussian optics can be attributed to a beam having infinitely low divergence, one can, according to the Huygens-Fresnel principle, only capture a diffraction pattern in the far-field caused by interference of wavelets emitted from different parts of the object (Fig. 2.8a, b). In DIH however, one expects a spherical wave, for sample illumination as well as for serving as reference, $U_{\mathrm{ref}}(\mathbf{r}) = \exp(\mathrm{i}\mathbf{kr})/r$ arriving at the object. Since the spherical pinhole wavefronts have a divergence similar to the object wavefronts, an interference pattern from the reference wave $U_{\mathrm{ref}}(\mathbf{r})$ and the scattered wave $U_{\mathrm{scat}}(\mathbf{r})$ forms in the detector plane (Fig. 2.8c). From the field $U_{\mathrm{det}}(\mathbf{r})$ in the detection plane one can only record the intensities, which can be written as

$$\begin{aligned} I(\mathbf{r}) = |U_{\mathrm{det}}(\mathbf{r})|^2 &= |U_{\mathrm{ref}}(\mathbf{r}) + U_{\mathrm{scat}}(\mathbf{r})|^2 \\ &= [U^*_{\mathrm{ref}}(\mathbf{r})U_{\mathrm{scat}}(\mathbf{r}) + U_{\mathrm{ref}}(\mathbf{r})U^*_{\mathrm{scat}}(\mathbf{r})] \\ &\quad + |U_{\mathrm{scat}}(\mathbf{r})|^2 + |U_{\mathrm{ref}}(\mathbf{r})|^2. \end{aligned} \tag{2.43}$$

Inspecting the terms in Eq. 2.43 one finds the interference between the scattered and the reference wave in the first term (in the brackets), which gives rise to the hologram (Fig. 2.8d). The diffraction pattern analogue to CDI can be found in the second term. The third term is the far-field of the source, which, in case of a small point-like source, is just a constant field that could be subtracted. Hence for DIH it is essential that the object does not entirely block the reference wave. However, even for apertures DIH works, but the fringe contrast is getting worse the more of the reference is blocked by an opaque object [70]. Moreover, the geometry of DIH is dictated by the source size, i.e. the divergence of the reference beam. If a pinhole is used for illuminating an object care must be taken that the zeroth order of the produced Airy pattern[11] covers both the object and the detector in order to get a well resolved hologram. Comparing Fig. 2.8b, d reveals an important advantage of in-line holography, which is that in DIH no beam block is needed to suppress the central high intensity maximum as it occurs in CDI. Moreover the intensity distribution across the

[11] An *Airy pattern* or *Airy disk* is the diffraction pattern caused from a round aperture featuring a bright central maximum, i.e. zeroth diffraction order, surrounded by dark and bright rings, i.e. higher order diffraction terms.

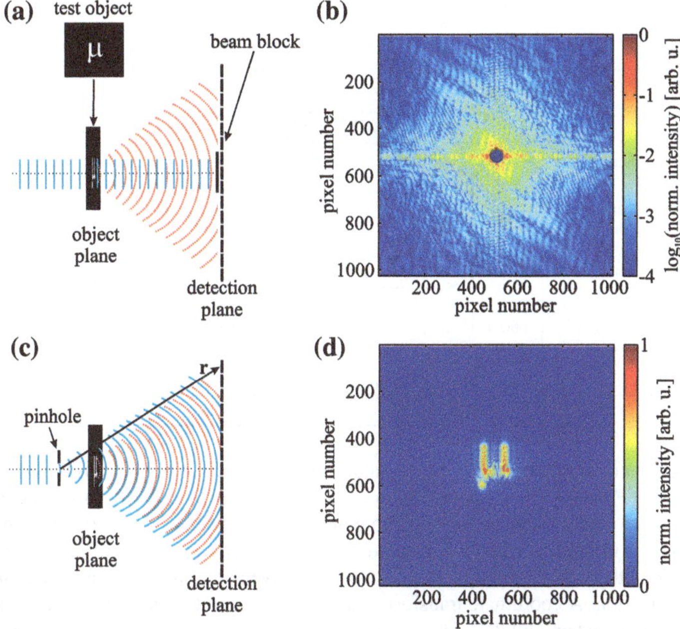

Fig. 2.8 Comparison of CDI and DIH with a μ-shaped aperture. **a** In a CDI configuration the object is illuminated with a plane wave (*solid blue line*), which can partly propagate through the aperture and partly gives rise to the diffracted wave (*red dotted line*). Through Babinet's principle one can also think of a μ-shaped object in free space, where part of the impinging wave far away from the object propagates undisturbed to the detector. Typically, a beam stop is used to suppress the bright central speckle in order to have a better use of the limited dynamic range of the detector. **b** The intensity of the diffraction pattern (logarithmic scale) caused by the object. **c** A pinhole is placed in front of the object in order to illuminate it with a spherical wave. Thus one has the reference wave (*solid blue line*) and the diffracted (object) wave (*dotted red line*) covering the detector causing interferences across the detector. This is a typical DIH setup. **d** A simulated hologram one would measure in the far-field of a 40 μm large μ-shaped aperture that is illuminated with a 1 μm pinhole 1.9 mm away with $\lambda = 38$ nm wavelength. Please note that the intensity scale in (**d**) is linear

detector is homogeneous compared to CDI diffraction patterns where the intensities of the fringes typically vary by several orders of magnitude across the detector. This is expressed by the logarithmic intensity scale in Fig. 2.8b and the linear intensity scale in Fig. 2.8d. Another consequence is that noise in the measurement of a hologram has much less effect compared to CDI, because small fluctuations contribute less to the overall result. A much more detailed theoretical analysis of DIH can for instance be found in the thesis from Schürmann [71].

In Chap. 3 the setup that was used for DIH in this work is introduced. In a nutshell it consists of a pinhole having a certain diameter that is placed in the focus of a monochromatized HHG beam. This pinhole generates the spherical illumination wave, which is then diffracted at the object. The undisturbed part of this wave propagates to the detector and further serves as reference wave. In the far-field a CCD records the hologram. An excellent book from Poon and Banerjee [72] covers

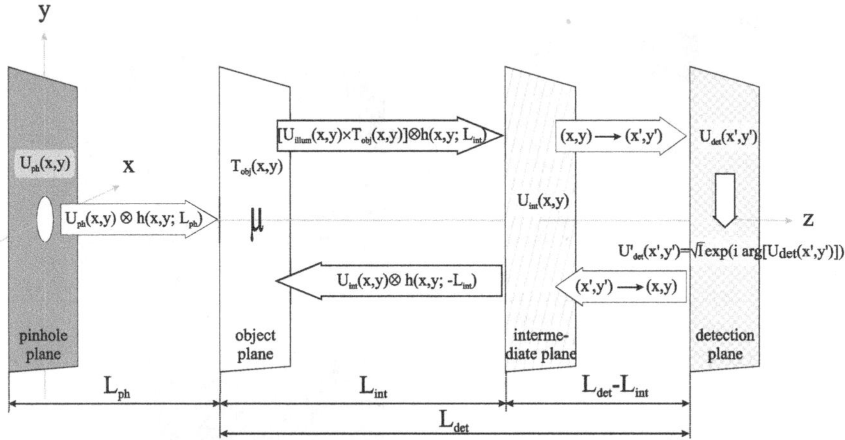

Fig. 2.9 Reconstruction scheme for the digital in-line holograms as used in this thesis. The convolution ($\otimes$) of the Fresnel propagator $h(x, y; z)$ with an input field $U(x, y)$ was used to propagate the field between the planes. Starting at the pinhole plane one propagates the field to the object plane, where the field is multiplied with an in general complex-valued transmission function of the object $T_{\text{obj}}(x, y)$. Then the field is propagated to the far-field given at an intermediate plane, see text for detailed explanation. A coordinate transform scales the field to the detector where the phases are kept and the measured amplitudes $\sqrt{I}$ are enforced. After back propagation to the object plane the object function is updated. These steps are repeated until convergence

this topic and its numerical integration in depth. Here, some of the notations given there will be used for the reconstruction. In order to reconstruct these holograms the following procedure is used (Fig. 2.9):

1. One starts in the pinhole plane with a field $U_{\text{ph}}(x, y)$ that has a flat phase and a radial distribution.
2. According to Chap. 3 in [72] one can compute the Fresnel diffraction pattern after a distance z (see Fig. 2.4), by convoluting the input field $U_{\text{ph}}(x, y)$ with the spatial impulse response $h(x, y; z)$

$$h(x, y; z) = \exp(-ikz)\frac{ik}{2\pi z}\exp\left[-ik\frac{(x^2 + y^2)}{2z}\right], \tag{2.44}$$

also known as the *Fresnel propagator*. The numerical implementation can be done by Fourier transforming the propagator h and the field U, multiplying them in Fourier space and Fourier transforming the result back, i.e.

$$U(x, y)\big|_z = \alpha\mathcal{F}^{-1}\{\mathcal{F}[h(x, y; z)] \times \mathcal{F}[U(x, y)]\}, \tag{2.45}$$

where α is a constant scale factor depending on the actual implementation of the FFT algorithm. Using this, the illumination field at the object plane can be computed by $U_{\text{illum}}(x, y) = U_{\text{ph}} \otimes h(x, y; L_{\text{ph}})$ when the pinhole plane and object plane are L_{ph} apart. This is essentially an Airy pattern. The experiment is aligned such that the zeroth order of the pattern, which contains 84 % of the energy, fits to the sample.

3. Now, in the object plane, one multiplies $U_{illum}(x, y)$ with the transmission function of the object $T_{obj}(x, y)$, which could e. g. be a simple step-function having the spatial shape of the object if one considers an opaque aperture. From this one gets the object plane exit wave $U_{obj}(x, y) = U_{illum}(x, y) \times T_{obj}(x, y)$.
4. $U_{obj}(x, y)$ is then propagated to the far-field, which is an intermediate plane $U_{int}(x, y)$ a couple of Fresnel lengths (Eq. 2.12) behind the object plane. Due to the large divergence of the beam the field spreads quickly spatially, which would demand a large computation grid to sufficiently sample amplitude and phase on the full propagation distance to the detector. Thus, the wave is just propagated a short distance L_{int} to be safely in the far-field regime. As explained earlier in Sect. 2.2.2 the diffraction pattern, or hologram in this case, only spreads spatially when propagating further but does not change its shape. Hence one can use a coordinate transformation ($x \rightarrow x'$, $y \rightarrow y'$) to transform the intermediate plane to the actual detector plane and get $U_{det}(x', y')$. The coordinate transform is applied in order to have sufficient sampling at the object plane, while keeping the computation space maintainable. This procedure is illustrated in more detail in [67].
5. In the detector plane, methods inspired by the HIO algorithm (Sect. 2.3) are used, i.e. keeping the phases of $U_{det}(x', y')$ and replace the amplitudes with the square root of the measured intensities.
6. Then the wave is transformed back to the intermediate plane, i.e. ($x' \rightarrow x$, $y' \rightarrow y$), and propagated back to the object plane to get a new $U'_{obj}(x, y)$. Now a new estimate of the object transmission $T'_{obj}(x, y) = U'_{obj}(x, y) - U_{illum}(x, y)$ is calculated.[12]

Steps 3–6 are now iteratively repeated until convergence occurs. It is worth noting that there is some similarity to the algorithm described in [67], with the differences mainly being that here a convolution of the Fresnel propagator is used instead of analytical formulas. This severely speeds up the computation, because, as outlined above, the convolution can be done by using FFTs, which in turn can be effectively parallelized. Moreover, here the full system is computed starting from the source producing an illumination wave. Hence, in contrast to the algorithm reported in [67], there is no limitation to real objects, since here $T_{obj}(x, y)$ could in general be complex-valued, e.g. for partly transparent objects that would change phase and amplitude of the illuminating wave. Further, it can be used to determine the illumination field in great detail if a known object, e.g. a rectangular aperture, is enforced and one loops through

[12] Other iterative methods reported in literature make use of the Gerchberg-Saxton algorithm [73] or use a support constraint to retrieve the hologram [74, 75]. The important difference is that here the full illumination field, which is well characterized by the pinhole being illuminated with a XUV beam having good coherence properties, is used at every step, in contrast to the algorithms reported in literature, which mostly switch between the object plane and the detector plane and disregard the illumination field.

all the steps. It also proved to work successfully for samples that are mostly consisting of strongly absorbing regions to use $T_{\text{obj}}(x, y)$ just as a binary function determined by the shrink-wrap method analogous to the support function S in the HIO algorithm and then calculate the complex-valued sample in the final iteration of the algorithm by

$$U_{\text{sample}} = U_{\text{int}}(x, y) \otimes h(x, y; -L_{\text{int}}) - U_{\text{illum}}. \tag{2.46}$$

The idea behind this is that all hard edges caused by opaque parts of the object cause intense fringes and likely dominate the hologram. Hence the rough structure of the object is enforced first and the partially transparent parts are recovered in the final step. In the results in Sect. 4.1, samples that justify this approach will be presented.

It is worth noting that the holographic procedure offers the possibility of reconstructing the object in an arbitrary plane using one hologram. Thus in principle three-dimensional information about the object can be retrieved to a certain extent by refocussing to different planes [70]. Moreover, the quality of the reconstruction can be enhanced by measuring the amplitude of the reference wave and then using this as an illumination field instead of the numerically produced illumination field discussed in this section.

The resolution, as was discussed in the previous sections, depends on the numerical aperture according to Eq. 1.1. For digital in-line holography using spherical waves the NA is [70]

$$\text{NA} = \frac{w}{2\sqrt{\left(\frac{w}{2}\right)^2 + L_{\text{det}}^2}}, \tag{2.47}$$

where w is the width of the detector. The NA is obviously independent of the source size, however, as discussed earlier in this section one needs a homogeneous illumination by employing the central maximum of the Airy pattern produced by the pinhole. The angular radius θ_0 of the first dark ring of an Airy pattern is given by

$$\sin\theta_0 = 1.22\frac{\lambda}{a}, \tag{2.48}$$

where a is the diameter of the aperture. The small angle approximation $\sin\theta_0 \approx \tan\theta_0$ is almost always justified in DIH experiments. Hence, one can calculate the approximate diameter of the central maximum of the Airy disk on the detector placed at L_{det}[13]

$$w_{\text{Airy}} = 2 \cdot 1.22\frac{\lambda}{a}L_{\text{det}}. \tag{2.49}$$

Physically this means that one observes a larger central maximum for a smaller pinhole if the detector is kept at constant distance. If $w_{\text{Airy}} < w$ and if the higher order

[13] More precisely it would be $L_{\text{det}} + L_{\text{ph}}$, however, for all experiments presented in this thesis $L_{\text{det}} \gg L_{\text{ph}}$, hence simply L_{det} is used.

maxima of the Airy pattern are neglected one can calculate an effective numerical aperture of the system

$$\mathrm{NA_{eff}} = \frac{1.22\frac{\lambda}{a}L_{\mathrm{det}}}{\sqrt{\left(1.22\frac{\lambda}{a}L_{\mathrm{det}}\right)^2 + L_{\mathrm{det}}^2}} = \frac{1}{\sqrt{1+0.67\frac{a^2}{\lambda^2}}}, \tag{2.50}$$

which now is obviously only dependent on the pinhole size and the wavelength. For XUV DIH usually $a \gg \lambda$, which allows to simplify Eq. 2.50 to $\mathrm{NA_{eff}} \approx \frac{1.22\lambda}{a}$. The practical implication of this result is that one should match the pinhole size to the dimensions of the detector and its anticipated distance [76]. In this case one can compute the NA and thus the achievable resolution simply by Eq. 2.47. Of course one could argue to use a pinhole so tiny that $w_{\mathrm{Airy}} \gg w$ is always fulfilled. However, in table-top HHG experiments the XUV flux is not excessive and thus best use of the available photons should be made. Hence, it can be concluded that it is beneficial to choose a pinhole that fits the detector and its achievable distance from the pinhole in order to realize the best performance in terms of the photon flux and the achievable resolution. In view of maximizing the magnification ζ

$$\zeta = L_{\mathrm{det}}/L_{\mathrm{ph}}, \tag{2.51}$$

which is inferred from the intercept theorem, one has to place the sample as close to the pinhole as mechanically possible.[14] On the other hand the field of view is minimized in this case. If the distance between sample and pinhole is too large, one cannot resolve the fringes if the Shannon-Nyquist theorem is violated. Hence, the optimal position of the sample relative to the pinhole and the CCD is somewhere in between and should be set such that the period of the smallest fringes is in the range of e.g. 5–10 pixels on the CCD [77]. In that case the FOV and object space NA, and thus the achievable resolution, are optimal.

If, however, $L_{\mathrm{det}} \gg L_{\mathrm{ph}}$, i.e. when the object is close to the pinhole, one could think of a spherical particle in the object plane at approximately the distance of the pinhole. To have a resolvable fringe one needs to have at least the first ring of the Airy pattern produced by the particle on the detector,[15] hence the discussion is exactly the same as above for the pinhole. For example, if one thinks of a 1 μm pinhole as the reference source and a round particle of e.g. 500 nm diameter in the object plane one will not observe any fringe from that particle on the detector because the emission cone of its partial wave is coarser than the one from the pinhole. In terms of the back projection this means that one cannot resolve this particle. The conclusion from this discussion is that the size of the pinhole is not just limiting the effective NA but also the resolution itself. As a rule of thumb it can be summarized that a resolution of approximately the size of the pinhole is achievable.

---

[14] For typical XUV wavelengths and pinhole sizes in the order of a micron the Fraunhofer condition (Eq. 2.12) is already fulfilled a few tens of microns behind the pinhole. Hence, the Fraunhofer condition is not limiting in XUV DIH.

[15] This is analogous to the Rayleigh criterion.

2.5 CDI State-of-the-Art and Chapter Summary

The CDI technique presented in the fundamentals chapter allows imaging of isolated objects at the nanometer scale without the use of imaging optics. This renders those systems very compact. In the case of table-top CDI experiments, the whole apparatus can fit on an optical table of a few square meters. Another major benefit is that lossy optical elements, such as Fresnel zone plates, are avoided. Moreover, the mechanical stability of the imaging system is less important, because lateral shifts of the object would only affect the phase in the far-field (Fourier space), which in any case cannot be measured. This is quite astonishing if one considers achieving a few nanometers of resolution while the mechanical stability of the system only needs to ensure that the beam homogeneously overlaps with the object during exposure. If one would use imaging optics instead, one would need a mechanical stability comparable to the desired resolution, which can be very challenging if nanometer resolution is aimed for. One major drawback of CDI, however, is that so far only isolated objects were considered, which could be e.g. an isolated cell, a structured aperture or a sample on a fixed support frame. It was demonstrated that digital in-line holography can be used to solve this problem at the expense of introducing another reference wave forming object, e.g. a pinhole. The latter will typically limit the achievable flux of the source and dictate the geometry of the imaging system. Moreover, the mechanical stability issue described above arises.

In recent years a technique called *ptychography* was developed, which overcomes this drawback [78, 79]. Ptychography uses a set of diffraction patterns obtained from an extended object, where the illuminated region is shifted from one exposure to the next while each exposure region overlaps with the previous. This produces a set of overdetermined diffraction patterns, which can then again be used to reconstruct the whole object plane. An extension of ptychography is called *keyhole diffraction imaging* [80] which is a combination of curved beam illumination at high numerical apertures, i.e. in-line holography combined with CDI, with ptychographic scanning. The latter technique shows quicker convergence behavior compared to pure ptychography. It is worth noting that the drawback of these techniques is again the need for high mechanical stability and limitations imposed by tight focusing optics, which currently restricts the application to synchrotrons and free-electron laser sources.[16] However, very recently first results on ptychography using table-top sources were published [81, 82].

Over the last decade the general field of CDI developed rapidly at synchrotrons and later at free-electron lasers. From its first experimental demonstration [83] it went on to image biological specimens such as single cells [84, 85] at a resolution of a few nanometers. An important improvement was introduced by Raines and co-workers who found that one can even extract three-dimensional data from a single diffraction pattern [86] if it is captured at a high numerical aperture. Other realizations of full 3D imaging combine tomographic techniques with CDI [28, 87]. The next goal

[16] This is because the high flux of such sources allows extremely short exposures down to single shot measurements which circumvents the stability problem.

that is to be achieved with novel free-electron laser sources, such as the EUROPEAN XFEL project, is atomic resolution, i.e. $\Delta r < 0.1$ nm. The major problem faced here is that the photon energies necessary to achieve this resolution in principle are so high that matter is destroyed already after one exposure [88–90]. Hence, only single shot exposures can be used. However, for atomic resolution $> 10^{14}$ photons/μm^2 in a single pulse would be necessary to get a sufficient signal in the diffraction pattern [91]. These numbers are out of reach for current FEL machines. Hence atomic resolution is one of the major goals for the coming years in CDI. In the meantime averaging over diffraction patterns obtained from replicas of the same objects was proposed [92] and demonstrated [93, 94] for proteins. Additional contrast is obtained by growing the protein as a crystal to get a macroscopic diffraction pattern of the same protein. Molecular reconstruction techniques are then used to fit the molecular structure to the diffraction pattern [93]. These techniques include the full three-dimensional reconstruction of the object.

Broad usage of CDI in all fields of science and especially for medical application is not feasible at next-generation synchrotrons and free-electron lasers, because such large scale facilities have limited access and are expensive to operate. An alternative are HHG sources that offer excellent XUV beams [13, 14] in terms of coherence allowing for table-top CDI experiments. Limiting is the flux in terms of exposure time and the bandwidth of the harmonics (cf. Eq. 2.29), and the achievable wavelengths in terms of resolution. Table-top CDI was first demonstrated by Sandberg et al. [95], while first holography measurements were done by Bartels et al. [13]. Abbey et al. demonstrated that the limitation due to bandwidth can be overcome for sufficiently discrete structures [96]. Likewise this can be applied to the harmonics frequency comb [97], while one has to keep in mind that this may work only for discrete objects that generate a limited amount of fringes such that the fringes caused by different harmonics are not overlapping. However, if this condition is fulfilled one can take advantage of the improved overall flux. Using more powerful laser sources even single shot HHG imaging was demonstrated [98, 99]. The flux from typical HHG sources does not endanger the detector unlike at FELs, hence the detector can be placed very close to the sample to achieve a high numerical aperture to take full advantage of the short wavelength radiation. Using a high numerical aperture in a table-top XUV experiment a resolution of 22 nm was achieved using 13.5 nm wavelength [100] which equals 1.6λ. It is also worth mentioning that HHG offers very short XUV bursts in the range of attoseconds to femtoseconds with the availability of a time-locked infrared laser, thus offering perfect capabilities for time resolved measurements [101, 102].

References

1. Boyd, R.W.: Nonlinear Optics. Academic Press, London (2008)
2. Schaefer, C., Niedrig, H., Bergmann, L., Eichler, H.-J.: Lehrbuch der Experimentalphysik. Walter de Gruyter, Berlin (2004) (Optik)

3. McPherson, A., Gibson, G., Jara, H., Johann, U., Luk, T.S., Mcintyre, I.A., Boyer, K., Rhodes, C.K.: Studies of multiphoton production of vacuum ultraviolet-radiation in the rare-gases. J. Opt. Soc. Am. B **4**(4), 595–601 (1987)
4. Li, X.F., L'Huillier, A., Ferray, M., Lompre, L.A., Mainfray, G.: Multiple-harmonic generation in rare gases at high laser intensity. Phys. Rev. A **39**(11), 5751–5761 (1989)
5. Seres, J., Seres, E., Verhoef, A.J., Tempea, G., Streli, C., Wobrauschek, P., Yakovlev, V., Scrinzi, A., Spielmann, C., Krausz, F.: Laser technology: source of coherent kiloelectronvolt X-rays. Nature **433**(7026), 596–596 (2005)
6. Popmintchev, T., Chen, M.C., Popmintchev, D., Arpin, P., Brown, S., Alisauskas, S., Andriukaitis, G., Balciunas, T., Mucke, O.D., Pugzlys, A., Baltuska, A., Shim, B., Schrauth, S.E., Gaeta, A., Hernandez-Garcia, C., Plaja, L., Becker, A., Jaron-Becker, A., Murnane, M.M., Kapteyn, H.C.: Bright coherent ultrahigh harmonics in the keV X-ray regime from mid-infrared femtosecond lasers. Science **336**(6086), 1287–1291 (2012)
7. Krause, J.L., Schafer, K.J., Kulander, K.C.: High-order harmonic generation from atoms and ions in the high intensity regime. Phys. Rev. Lett. **68**(24), 3535–3538 (1992)
8. Corkum, P.B.: Plasma perspective on strong field multiphoton ionization. Phys. Rev. Lett. **71**(13), 1994–1997 (1993)
9. Awasthi, M., Vanne, Y.V., Saenz, A., Castro, A., Decleva, P.: Single-active-electron approximation for describing molecules in ultrashort laser pulses and its application to molecular hydrogen. Phys. Rev. A **77**(6), 063403 (2008)
10. Keldysh, L.V.: Ionization in field of a strong electromagnetic wave. Sov. Phys. JETP **20**(5), 1307–1314 (1965)
11. Brabec, T., Krausz, F.: Intense few-cycle laser fields: frontiers of nonlinear optics. Rev. Mod. Phys. **72**(2), 545–591 (2000)
12. Lewenstein, M., Balcou, Ph, Ivanov, M.Y., L'Huillier, A., Corkum, P.B.: Theory of high-harmonic generation by low-frequency laser fields. Phys. Rev. A **49**(3), 2117–2132 (1994)
13. Bartels, R.A., Paul, A., Green, H., Kapteyn, H.C., Murnane, M.M., Backus, S., Christov, I.P., Liu, Y.W., Attwood, D., Jacobsen, C.: Generation of spatially coherent light at extreme ultraviolet wavelengths. Science **297**(5580), 376–378 (2002)
14. Ditmire, T., Gumbrell, E.T., Smith, R.A., Tisch, J.W.G., Meyerhofer, D.D., Hutchinson, M.H.R.: Spatial coherence measurement of soft X-ray radiation produced by high order harmonic generation. Phys. Rev. Lett. **77**(23), 4756–4759 (1996)
15. Constant, E., Garzella, D., Breger, P., Mevel, E., Dorrer, C., Le Blanc, C., Salin, F., Agostini, P.: Optimizing high harmonic generation in absorbing gases: model and experiment. Phys. Rev. Lett. **82**(8), 1668–1671 (1999)
16. Takahashi, E., Nabekawa, Y., Midorikawa, K.: Generation of 10-μJ coherent extreme-ultraviolet light by use of high-order harmonics. Opt. Lett. **27**(21), 1920–1922 (2002)
17. Popmintchev, T., Chen, M.C., Bahabad, A., Gerrity, M., Sidorenko, P., Cohen, O., Christov, I.P., Murnane, M.M., Kapteyn, H.C.: Phase matching of high harmonic generation in the soft and hard X-ray regions of the spectrum. Proc. Natl. Acad. Sci. USA **106**(26), 10516–10521 (2009)
18. Paul, P.M., Toma, E.S., Breger, P., Mullot, G., Auge, F., Balcou, Ph., Muller, H.G., Agostini, P.: Observation of a train of attosecond pulses from high harmonic generation. Science **292**(5522), 1689–1692 (2001)
19. Myers, O.E.: Studies of transmission zone plates. Am. J. Phys. **19**(6), 359–365 (1951)
20. Horowitz, P.: Scanning X-ray microscope using synchrotron radiation. Science **178**(4061), 608–611 (1972)
21. Lee, K.H., Park, S.B., Singhal, H., Nam, C.H.: Ultrafast direct imaging using a single high harmonic burst. Opt. Lett. **38**(8), 1253–1255 (2013)
22. Attwood, D.: Soft X-rays and Extreme Ultraviolet Radiation: Principles and Applications. Cambridge University Press, Cambridge (2007)
23. Jackson, J.D., Witte, C., Mueller, K.: Klassische Elektrodynamik, vol. 3. Walter de Gruyter, Berlin (2006)
24. Gu, M.: Advanced Optical Imaging Theory. Springer, Berlin (2000)

25. Williams, G.J., Quiney, H.M., Dhal, B.B., Tran, C.Q., Nugent, K.A., Peele, A.G., Paterson, D., de Jonge, M.D.: Fresnel coherent diffractive imaging. Phys. Rev. Lett. **97**(2), 025506 (2006)
26. Consortini, A.: Trends in Optics: Research, Developments, and Applications. Academic Press, London (1996)
27. Ewald, P.P.: Zur Theorie der Interferenzen der Roentgenstrahlen in Kristallen. Phys. Zeitschrift **14**, 465–472 (1913)
28. Chapman, H.N., Barty, A., Marchesini, S., Noy, A., Hau-Riege, S.P., Cui, C., Howells, M.R., Rosen, R., He, H., Spence, J.C., Weierstall, U., Beetz, T., Jacobsen, C., Shapiro, D.: High-resolution ab initio three-dimensional X-ray diffraction microscopy. J. Opt. Soc. Am. A Opt. Image Sci. Vis. **23**(5), 1179–1200 (2006)
29. Sayre, D.: Some Implications of a theorem due to shannon. Acta Crystallogr. **5**(6), 843–843 (1952)
30. Shannon, C.E.: Communication in the presence of noise. Proc. IRE **37**(1), 10–21 (1949)
31. Sayre, D.: Prospects for long-wavelength X-ray microscopy and diffraction. In: Schlenker, M., Fink, M., Goedgebuer, J.P., Malgrange, C., Vieenot, J.C., Wade, R.H. (eds.) Imaging Processes and Coherence in Physics, vol. 112, pp. 229–235. Springer, Berlin (1980)
32. Bates, R.H.T.: Fourier phase problems are uniquely solvable in more than one dimension. I: underlying theory. Optik **61**, 247–262 (1982)
33. Miao, J., Sayre, D.: On possible extensions of X-ray crystallography through diffraction-pattern oversampling. Acta Crystallogr. A **56**(6), 596–605 (2000)
34. Miao, J., Ishikawa, T., Anderson, E.H., Hodgson, K.O.: Phase retrieval of diffraction patterns from noncrystalline samples using the oversampling method. Phys. Rev. B **67**(17), 174104 (2003)
35. Spence, J.C., Weierstall, U., Howells, M.: Coherence and sampling requirements for diffractive imaging. Ultramicroscopy **101**(2–4), 149–152 (2004)
36. Whitehead, L.W., Williams, G.J., Quiney, H.M., Vine, D.J., Dilanian, R.A., Flewett, S., Nugent, K.A., Peele, A.G. Balaur, E. McNulty, I.: Diffractive imaging using partially coherent X-rays. Phys. Rev. Lett. **103**(24), 243902 (2009)
37. Gerchberg, R.W., Saxton, W.O.: Phase determination from image and diffraction plane pictures in electron-microscope. Optik **34**(3), 275–283 (1971)
38. Fienup, J.R.: Phase retrieval algorithms: a comparison. Appl. Opt. **21**(15), 2758–2769 (1982)
39. Fienup, J.R., Crimmins, T.R., Holsztynski, W.: Reconstruction of the support of an object from the support of its auto-correlation. J. Opt. Soc. Am. **72**(5), 610–624 (1982)
40. Marchesini, S., Chapman, H.N., Hau-Riege, S.P., London, R.A., Szoke, A., He, H., Howells, M.R., Padmore, H., Rosen, R., Spence, J.C.H., Weierstall, U.: Coherent X-ray diffractive imaging: applications and limitations. Opt. Express **11**(19), 2344–2353 (2003)
41. Martin, A.V., Wang, F., Loh, N.D., Ekeberg, T., Maia, F.R.N.C., Hantke, M., van der Schot, G., Hampton, C.Y., Sierra, R.G., Aquila, A., Bajt, S., Barthelmess, M., Bostedt, C., Bozek, J.D., Coppola, N., Epp, S.W., Erk, B., Fleckenstein, H., Foucar, L., Frank, M., Graafsma, H., Gumprecht, L., Hartmann, A., Hartmann, R., Hauser, G., Hirsemann, H., Holl, P., Kassemeyer, S., Kimmel, N., Liang, M., Lomb, L., Marchesini, S., Nass, K., Pedersoli, E., Reich, C., Rolles, D., Rudek, B., Rudenko, A., Schulz, J., Shoeman, R.L., Soltau, H., Starodub, D., Steinbrener, J., Stellato, F., Strueder, L., Ullrich, J., Weidenspointner, G., White, T.A., Wunderer, C.B., Barty, A., Schlichting, I., Bogan, M.J., Chapman, H.N.: Noise-robust coherent diffractive imaging with a single diffraction pattern. Opt. Express **20**(15), 16650–16661 (2012)
42. Luke, D.R.: Relaxed averaged alternating reflections for diffraction imaging. Inverse Prob. **21**(1), 37–50 (2005)
43. Marchesini, S., He, H., Chapman, H.N., Hau-Riege, S.P., Noy, A., Howells, M.R., Weierstall, U., Spence, J.C.H.: X-ray image reconstruction from a diffraction pattern alone. Phys. Rev. B **68**(14), 140101 (2003)
44. Baeck, T., Schwefel, H.-P.: An overview of evolutionary algorithms for parameter optimization. Evol. Comput. **1**(1), 1–23 (1993)

45. Miao, J., Chen, C.-C., Song, C., Nishino, Y., Kohmura, Y., Ishikawa, T., Ramunno-Johnson, D., Lee, T.-K., Risbud, S.H.: Three-dimensional GaN-Ga_2O_3 core shell structure revealed by X-ray diffraction microscopy. Phys. Rev. Lett. **97**(21), 215503 (2006)
46. Miao, J., Sayre, D., Chapman, H.N.: Phase retrieval from the magnitude of the Fourier transforms of nonperiodic objects. J. Opt. Soc. Am. A **15**(6), 1662–1669 (1998)
47. Gabor, D.: Microscopy by reconstructed wave-fronts. Proc. R. Soc. Lon. Ser. A **197**(1051), 454–487 (1949)
48. Toal, V.: Introduction to Holography. CRC Press, Boca Raton (2011)
49. Howells, M., Jacobsen, C., Kirz, J., Feder, R., Mcquaid, K., Rothman, S.: X-ray holograms at improved resolution—a study of Zymogen Granules. Science **238**(4826), 514–517 (1987)
50. McNulty, I., Kirz, J., Jacobsen, C., Anderson, E.H., Howells, M.R., Kern, D.P.: High-resolution imaging by Fourier transform X-ray holography. Science **256**(5059), 1009–1012 (1992)
51. Eisebitt, S., Luning, J., Schlotter, W.F., Lorgen, M., Hellwig, O., Eberhardt, W., Stohr, J.: Lensless imaging of magnetic nanostructures by X-ray spectro-holography. Nature **432**(7019), 885–888 (2004)
52. Chapman, H.N., Hau-Riege, S.P., Bogan, M.J., Bajt, S., Barty, A., Boutet, S., Marchesini, S., Frank, M., Woods, B.W., Benner, W.H., London, R.A., Rohner, U., Szoke, A., Spiller, E., Moller, T., Bostedt, C., Shapiro, D.A., Kuhlmann, M., Treusch, R., Plonjes, E., Burmeister, F., Bergh, M., Caleman, C., Huldt, G., Seibert, M.M., Hajdu, J.: Femtosecond time-delay X-ray holography. Nature **448**(7154), 676–679 (2007)
53. Quiney, H.M., Peele, A.G., Cai, Z., Paterson, D., Nugent, K.A.: Diffractive imaging of highly focused X-ray fields. Nat. Phys. **2**(2), 101–104 (2006)
54. Schropp, A., Hoppe, R., Meier, V., Patommel, J., Seiboth, F., Lee, H.J., Nagler, B., Galtier, E.C., Arnold, B., Zastrau, U., Hastings, J.B., Nilsson, D., Uhlen, F., Vogt, U., Hertz, H.M., Schroer, C.G.: Full spatial characterization of a nanofocused X-ray free-electron laser beam by ptychographic imaging. Sci. Rep. **3**, 1633 (2013)
55. Sandberg, R.L., Raymondson, D.A., La, O.V.C., Paul, A., Raines, K.S., Miao, J., Murnane, M.M., Kapteyn, H.C., Schlotter, W.F.: Tabletop soft-X-ray Fourier transform holography with 50 nm resolution. Opt. Lett. **34**(11), 1618–1620 (2009)
56. Williams, G.J., Quiney, H.M., Dahl, B.B., Tran, C.Q., Peele, A.G., Nugent, K.A., De Jonge, M.D., Paterson, D.: Curved beam coherent diffractive imaging. Thin Solid Films **515**(14), 5553–5556 (2007)
57. Williams, G.J., Quiney, H.M., Peele, A.G., Nugent, K.A.: Fresnel coherent diffractive imaging: treatment and analysis of data. New J. Phys. **12**, 035020 (2010)
58. Geilhufe, J., Pfau, B., Schneider, M., Buttner, F., Gunther, C.M., Werner, S., Schaffert, S., Guehrs, E., Frommel, S., Klaui, M., Eisebitt, S.: Monolithic focused reference beam X-ray holography. Nat. Commun. **5**, 3008 (2014)
59. Schlotter, W.F., Rick, R., Chen, K., Scherz, A., Stohr, J., Luning, J., Eisebitt, S., Gunther, C., Eberhardt, W., Hellwig, O., McNulty, I.: Multiple reference Fourier transform holography with soft X-rays. Appl. Phys. Lett. **89**(16), 163112 (2006)
60. Gunther, C.M., Pfau, B., Mitzner, R., Siemer, B., Roling, S., Zacharias, H., Kutz, O., Rudolph, I., Schondelmaier, D., Treusch, R., Eisebitt, S.: Sequential femtosecond X-ray imaging. Nat. Photonics **5**(2), 99–102 (2011)
61. Guizar-Sicairos, M., Fienup, J.R.: Holography with extended reference by autocorrelation linear differential operation. Opt. Express **15**(26), 17592–17612 (2007)
62. Gauthier, D., Guizar-Sicairos, M., Ge, X., Boutu, W., Carre, B., Fienup, J.R., Merdji, H.: Single-shot Femtosecond X-ray holography using extended references. Phys. Rev. Lett. **105**(9), 093901 (2010)
63. Zhu, D., Guizar-Sicairos, M., Wu, B., Scherz, A., Acremann, Y., Tyliszczak, T., Fischer, P., Friedenberger, N., Ollefs, K., Farle, M., Fienup, J.R., Stohr, J.: High-resolution X-ray lensless imaging by differential holographic encoding. Phys. Rev. Lett. **105**(4), 043901 (2010)
64. Wachulak, P.W., Marconi, M.C., Bartels, R.A., Menoni, C.S., Rocca, J.J.: Holographic imaging with a nanometer resolution using compact table-top EUV laser. Opto-Electron. Rev. **18**(1), 80–90 (2010)

65. Schnars, U., Jueptner, W.: Digital Holography: Digital Hologram Recording, Numerical Reconstruction, and Related Techniques. Springer, Berlin (2005)
66. Morlens, A.S., Gautier, J., Rey, G., Zeitoun, P., Caumes, J.P., Kos-Rosset, M., Merdji, H., Kazamias, S., Casson, K., Fajardo, M.: Submicrometer digital in-line holographic microscopy at 32 nm with high-order harmonics. Opt. Lett. **31**(21), 3095–3097 (2006)
67. Genoud, G., Guilbaud, O., Mengotti, E., Pettersson, S.G., Georgiadou, E., Pourtal, E., Wahlstroem, C.G., L'Huillier, A.: XUV digital in-line holography using high-order harmonics. Appl. Phys. B **90**(3–4), 533–538 (2008)
68. Schnars, U., Juptner, W.P.: Digital recording and reconstruction of holograms in hologram interferometry and shearography. Appl. Opt. **33**(20), 4373–4377 (1994)
69. Schnars, U., Jueptner, W.: Direct recording of holograms by a CCD target and numerical reconstruction. Appl. Opt. **33**(2), 179–181 (1994)
70. Garcia-Sucerquia, J., Xu, W., Jericho, S.K., Klages, P., Jericho, M.H., Kreuzer, H.J.: Digital in-line holographic microscopy. Appl. Opt. **45**(5), 836–850 (2006)
71. Schuermann, M.: Digital in-line holographic microscopy with various wavelengths and point sources applied to static and fluidic specimens. Ph.D. thesis, University of Heidelberg (2007)
72. Poon, T.C., Banerjee, P.P.: Contemporary Optical Image Processing with MATLAB. Elsevier Science, Oxford (2001)
73. Liu, G., Scott, P.D.: Phase retrieval and twin-image elimination for in-line Fresnel holograms. J. Opt. Soc. Am. A **4**(1), 159–165 (1987)
74. Koren, G., Polack, F., Joyeux, D.: Iterative algorithms for twin-image elimination in in-line holography using finite-support constraints. J. Opt. Soc. Am. A **10**(3), 423–433 (1993)
75. Rong, L., Li, Y., Liu, S., Xiao, W., Pan, F., Wang, D.Y.: Iterative solution to twin image problem in in-line digital holography. Opt. Laser. Eng. **51**(5), 553–559 (2013)
76. Kreuzer, H.J., Jericho, M.J., Meinertzhagen, I.A., Xu, W.B.: Digital in-line holography with photons and electrons. J. Phys. Condens. Matter **13**(47), 10729–10741 (2001)
77. Kanka, M.: Bildrekonstruktion in der digitalen inline-holografischen Mikroskopie. Universitaetsverlag Ilmenau, Ilmenau (2011)
78. Thibault, P., Dierolf, M., Menzel, A., Bunk, O., David, C., Pfeiffer, F.: High-resolution scanning X-ray diffraction microscopy. Science **321**(5887), 379–382 (2008)
79. Thibault, P., Menzel, A.: Reconstructing state mixtures from diffraction measurements. Nature **494**(7435), 68–71 (2013)
80. Abbey, B., Nugent, K.A., Williams, G.J., Clark, J.N., Peele, A.G., Pfeifer, M.A., De Jonge, M., McNulty, I.: Keyhole coherent diffractive imaging. Nat. Phys. **4**(5), 394–398 (2008)
81. Seaberg, M.D., Zhang, F., Gardner, D.F., Shannon, C.E., Murnane, M.M., Kapteyn, H.C., Adams, D.E.: Tabletop nanometer extreme ultraviolet imaging in an extended reflection mode using coherent fresnel ptychography, pp. 1–9. arXiv: 1312.2049 (2013)
82. Zhang, B., Seaberg, M.D., Adams, D.E., Gardner, D.F., Shanblatt, E.R., Shaw, J.M., Chao, W., Gullikson, E.M., Salmassi, F., Kapteyn, H.C., Murnane, M.M.: Full field tabletop EUV coherent diffractive imaging in a transmission geometry. Opt. Express **21**(19), 21970–21980 (2013)
83. Miao, J.W., Charalambous, P., Kirz, J., Sayre, D.: Extending the methodology of X-ray crystallography to allow imaging of micrometre-sized non-crystalline specimens. Nature **400**(6742), 342–344 (1999)
84. Shapiro, D., Thibault, P., Beetz, T., Elser, V., Howells, M., Jacobsen, C., Kirz, J., Lima, E., Miao, H., Neiman, A.M., Sayre, D.: Biological imaging by soft X-ray diffraction microscopy. Proc. Natl. Acad. Sci. USA **102**(43), 15343–15346 (2005)
85. Huang, X., Nelson, J., Kirz, J., Lima, E., Marchesini, S., Miao, H., Neiman, A.M., Shapiro, D., Steinbrener, J., Stewart, A., Turner, J.J., Jacobsen, C.: Soft X-ray diffraction microscopy of a frozen hydrated yeast cell. Phys. Rev. Lett. **103**(19), 198101 (2009)
86. Raines, K.S., Salha, S., Sandberg, R.L., Jiang, H., Rodriguez, J.A., Fahimian, B.P., Kapteyn, H.C., Du, J., Miao, J.: Three-dimensional structure determination from a single view. Nature **463**(7278), 214–217 (2010)

87. Jiang, H., Song, C., Chen, C.C., Xu, R., Raines, K.S., Fahimian, B.P., Lu, C.H., Lee, T.K., Nakashima, A., Urano, J., Ishikawa, T., Tamanoi, F., Miao, J.: Quantitative 3D imaging of whole, unstained cells by using X-ray diffraction microscopy. Proc. Natl. Acad. Sci. USA **107**(25), 11234–11239 (2010)
88. Chapman, H.N., Barty, A., Bogan, M.J., Boutet, S., Frank, M., Hau-Riege, S.P., Marchesini, S., Woods, B.W., Bajt, S., Benner, H., London, R.A., Plonjes, E., Kuhlmann, M., Treusch, R., Dusterer, S., Tschentscher, T., Schneider, J.R., Spiller, E., Moller, T., Bostedt, C., Hoener, M., Shapiro, D.A., Hodgson, K.O., Van der Spoel, D., Burmeister, F., Bergh, M., Caleman, C., Huldt, G., Seibert, M.M., Maia, F.R.N.C., Lee, R.W., Szoke, A., Timneanu, N., Hajdu, J.: Femtosecond diffractive imaging with a soft-X-ray free-electron laser. Nat. Phys. **2**(12), 839–843 (2006)
89. Axford, D., Owen, R.L., Aishima, J., Foadi, J., Morgan, A.W., Robinson, J.I., Nettleship, J.E., Owens, R.J., Moraes, I., Fry, E.E., Grimes, J.M., Harlos, K., Kotecha, A., Ren, J., Sutton, G., Walter, T.S., Stuart, D.I., Evans, G.: In situ macromolecular crystallography using microbeams. Acta Crystallogr. D Biol. Crystallogr. **68**(5), 592–600 (2012)
90. Ziaja, B., Chapman, H.N., Faeustlin, R., Hau-Riege, S., Jurek, Z., Martin, A.V., Toleikis, S., Wang, F., Weckert, E., Santra, R.: Limitations of coherent diffractive imaging of single objects due to their damage by intense X-ray radiation. New J. Phys. **14**, 115015 (2012)
91. Neutze, R., Wouts, R., van der Spoel, D., Weckert, E., Hajdu, J.: Potential for biomolecular imaging with femtosecond X-ray pulses. Nature **406**(6797), 752–757 (2000)
92. Miao, J., Hodgson, K.O., Sayre, D.: An approach to three-dimensional structures of biomolecules by using single-molecule diffraction images. Proc. Natl. Acad. Sci. USA **98**(12), 6641–6645 (2001)
93. Boutet, S., Lomb, L., Williams, G.J., Barends, T.R.M., Aquila, A., Doak, R.B., Weierstall, U., DePonte, D.P., Steinbrener, J., Shoeman, R.L., Messerschmidt, M., Barty, A., White, T.A., Kassemeyer, S., Kirian, R.A., Seibert, M.M., Montanez, P.A., Kenney, C., Herbst, R., Hart, P., Pines, J., Haller, G., Gruner, S.M., Philipp, H.T., Tate, M.W., Hromalik, M., Koerner, L.J., van Bakel, N., Morse, J., Ghonsalves, W., Arnlund, D., Bogan, M.J., Caleman, C., Fromme, R., Hampton, C.Y., Hunter, M.S., Johansson, L.C., Katona, G., Kupitz, C., Liang, M.N., Martin, A.V., Nass, K., Redecke, L., Stellato, F., Timneanu, N., Wang, D.J., Zatsepin, N.A., Schafer, D., Defever, J., Neutze, R., Fromme, P., Spence, J.C.H., Chapman, H.N., Schlichting, I.: High-resolution protein structure determination by serial femtosecond crystallography. Science **337**(6092), 362–364 (2012)
94. Barends, T.R., Foucar, L., Botha, S., Doak, R.B., Shoeman, R.L., Nass, K., Koglin, J.E., Williams, G.J., Boutet, S., Messerschmidt, M., Schlichting, I.: De novo protein crystal structure determination from X-ray free-electron laser data. Nature **505**(7482), 244–247 (2014)
95. Sandberg, R.L., Paul, A., Raymondson, D.A., Haedrich, S., Gaudiosi, D.M., Holtsnider, J., Tobey, R.I., Cohen, O., Murnane, M.M., Kapteyn, H.C., Song, C.G., Miao, J.W., Liu, Y.W., Salmassi, F.: Lensless diffractive imaging using tabletop coherent high-harmonic soft-X-ray beams. Phys. Rev. Lett. **99**(9), 098103 (2007)
96. Abbey, B., Whitehead, L.W., Quiney, H.M., Vine, D.J., Cadenazzi, G.A., Henderson, C.A., Nugent, K.A., Balaur, E., Putkunz, C.T., Peele, A.G., Williams, G.J., McNulty, I.: Lensless imaging using broadband X-ray sources. Nat. Photonics **5**(7), 420–424 (2011)
97. Chen, B., Dilanian, R.A., Teichmann, S., Abbey, B., Peele, A.G., Williams, G.J., Hannaford, P., Van Dao, L., Quiney, H.M., Nugent, K.A.: Multiple wavelength diffractive imaging. Phys. Rev. A **79**(2), 023809 (2009)
98. Ravasio, A., Gauthier, D., Maia, F.R., Billon, M., Caumes, J.P., Garzella, D., Geleoc, M., Gobert, O., Hergott, J.F., Pena, A.M., Perez, H., Carre, B., Bourhis, E., Gierak, J., Madouri, A., Mailly, D., Schiedt, B., Fajardo, M., Gautier, J., Zeitoun, P., Bucksbaum, P.H., Hajdu, J., Merdji, H.: Single-shot diffractive imaging with a table-top femtosecond soft X-ray laser-harmonics source. Phys. Rev. Lett. **103**(2), 028104 (2009)
99. Malm, E.B., Monserud, N.C., Brown, C.G., Wachulak, P.W., Xu, H., Balakrishnan, G., Chao, W., Anderson, E., Marconi, M.C.: Tabletop single-shot extreme ultraviolet Fourier transform holography of an extended object. Opt. Express **21**(8), 9959–9966 (2013)

100. Seaberg, M.D., Adams, D.E., Townsend, E.L., Raymondson, D.A., Schlotter, W.F., Liu, Y.W., Menoni, C.S., Rong, L., Chen, C.C., Miao, J.W., Kapteyn, H.C., Murnane, M.M.: Ultrahigh 22 nm resolution coherent diffractive imaging using a desktop 13 nm high harmonic source. Opt. Express **19**(23), 22470–22479 (2011)
101. Tobey, R.I., Siemens, M.E., Cohen, O., Murnane, M.M., Kapteyn, H.C., Nelson, K.A.: Ultrafast extreme ultraviolet holography: dynamic monitoring of surface deformation. Opt. Lett. **32**(3), 286–288 (2007)
102. Prell, J.S., Borja, L.J., Neumark, D.M., Leone, S.R.: Simulation of attosecond-resolved imaging of the plasmon electric field in metallic nanoparticles. Ann. Phys. (Berlin) **525**(1–2), 151–161 (2013)

Chapter 3
Experimental Setup

The implementation of coherent XUV imaging techniques at the lab scale is the major topic of this thesis. Many promising applications were introduced in the previous chapter. At the same time limitations and obstacles in view of generating a versatile tool for scientific and medical application on a daily basis were discussed.

In Sect. 3.1.1 the main HHG source driven by an ultrafast titanium-doped sapphire laser (Ti:Sa) is described. In the subsequent Sect. 3.1.2 another HHG source employing a fiber laser driven chirped pulse amplification (FCPA) system is introduced. The characteristics of the HHG radiation produced with both sources are mentioned in the corresponding section. In Sect. 3.2 the different setup schemes for the CDI measurements presented in this work are explained. Likewise, an introduction of the setup for digital in-line holography follows (Sect. 3.3). In the final Sect. 3.4 the preparation of the measured data for the phase retrieval is discussed.

3.1 Realization and Characteristics of the HHG Source

3.1.1 Ultrafast Ti:Sa Laser Based HHG Source

Most of the experiments presented in this thesis were conducted using a commercial ultrafast femtosecond laser system (FEMTOPOWER COMPACT PRO from FEMTOLASERS). This Ti:Sa based system is depicted in Fig. 3.1. It consists of a femtosecond oscillator which is followed by a multipass CPA amplifier [1]. After compression one usually gets 1,000 ultrafast light pulses per second with approximately 1 mJ pulse energy having less than 30 fs pulse duration (FWHM) each. For the characterization of these pulses an interferometric autocorrelator [2] and a frequency-resolved optical gating (FROG) device are employed [3].

For HHG these pulses are focused, by means of a curved mirror ($f = 400$ mm, $f/\# = 15$), onto a nickel tube that is housed inside a high vacuum chamber. A schematic of the setup is depicted in Fig. 3.2a. The nickel tube has a diameter of

M.W. Zürch, *High-Resolution Extreme Ultraviolet Microscopy*,
Springer Theses, DOI 10.1007/978-3-319-12388-2_3

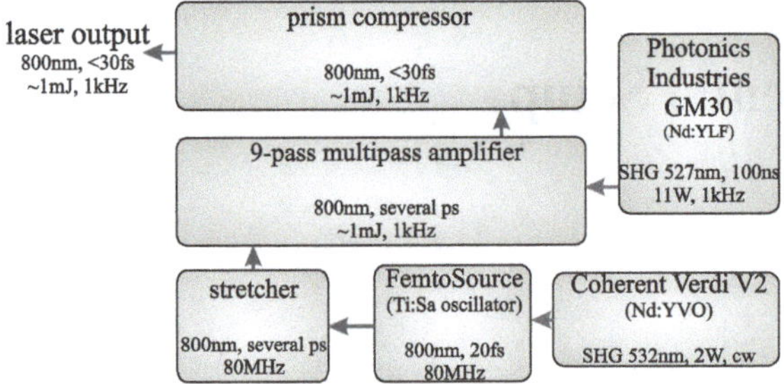

Fig. 3.1 Schematic of the commercial Ti:Sa CPA laser system used for most of the experiments presented in this thesis. Pulses from an ultrafast Ti:Sa oscillator are first temporally stretched, then amplified and subsequently compressed again using a prism compressor. The oscillator and the amplifier are pumped by a continuous wave (CW) frequency-doubled neodymium-doped yttrium orthovanadate (Nd:YVO) and a pulsed neodymium-doped yttrium lithium fluoride (Nd:YLF) laser respectively

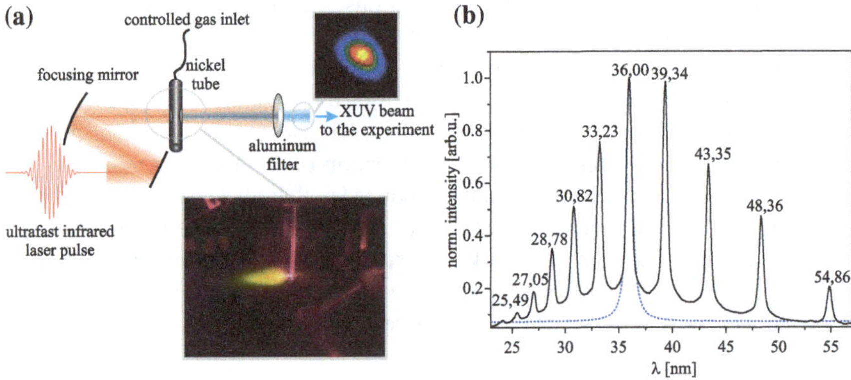

Fig. 3.2 Schematic and spectrum of HHG using a Ti:Sa driving laser. **a** The ultrafast laser pulse is focused into a gas nozzle consisting of a nickel tube. Through a controlled gas inlet a continuous stream of argon is maintained. In the interaction region (*bottom inset* photograph from above, laser comes from the left) HHG takes place. A subsequent aluminum filter suppresses the fundamental laser and only the harmonics (*top inset* measured beam profile of the harmonics in the far-field) can propagate to the experiment. **b** Typical HHG spectrum (*solid black line*) produced in argon; due to phase matching effects the harmonics around $\lambda = 36\,\text{nm}$ are the strongest. The cut-off is well below 25 nm. The wavelengths were tagged by fitting a Lorentzian function to each harmonic, example shown for $\lambda = 36\,\text{nm}$ by the *dotted blue line*

roughly 2 mm and is quenched and soldered at the end. A hole is drilled into the tube by the laser itself. From the open end of the nickel tube the gas is fed in with a defined backing pressure turning the tube into a gas cell. Using a beam profiler, the size of the focus can be estimated to approximately 50 μm (FWHM) diameter, which allows to estimate the achieved intensity level to $I = 2 \times 10^{14}\,\text{W/cm}^2$, which is sufficient to generate high harmonics (see Sect. 2.1). Phase matching [4] was maintained by shifting the laser focus relative to the nickel tube in the propagation direction (geometric phase matching) and by adjusting the gas pressure for maximized output. Such systems typically produce harmonics with μW average power per harmonic in the plateau region [5], which is also confirmed for the present source. For all experiments presented in this thesis and conducted with the Ti:Sa laser, argon with about 100 mbar backing pressure was employed. The exact pressure was adjusted on a daily basis by optimizing for highest flux at the desired wavelength. To suppress residual light of the much stronger infrared laser in the experimental chamber, thin aluminum foils ($d \approx 200$ nm to $d \approx 500$ nm depending on the experiment) were inserted downstream. The excellent spatial beam profile of the harmonics is depicted in the inset in Fig. 3.2a. For CDI, moreover, the spectral properties are important. A part of the typical spectrum produced by HHG in argon is depicted in Fig. 3.2b.[1] The central wavelengths of the harmonics were tagged by fitting Lorentzian functions to each harmonic. The exact wavelengths vary due to an ionization-induced blueshift [4] in the range of approximately 1 nm depending on the day-to-day performance of the laser. For coherent imaging purposes, as presented in this thesis, the harmonics having the highest flux, i.e. harmonics having a wavelength in the range $\lambda = 30$–40 nm, were used. From the Lorentzian fit—see the example dotted blue line in Fig. 3.2b—one can deduce the relative bandwidth, e.g. for the 23rd harmonic, to be

$$\left.\frac{\Delta\lambda}{\lambda}\right|_{\lambda=36\,\mathrm{nm}} = \frac{0.62 \pm 0.01\,\mathrm{nm}}{36\,\mathrm{nm}} \approx \frac{1}{58}. \tag{3.1}$$

That is a rather high value compared to synchrotrons where $\Delta\lambda/\lambda \approx 1/100000$ can be achieved [6]. Using this result and Eq. 2.29 one can now estimate the expected[2] resolution limit, which is induced by the limited temporal coherence, considering a single cell of 3 μm diameter as the object

$$\Delta r = \frac{Oa\Delta\lambda}{\lambda} = \frac{2 \times 3\,\mu\mathrm{m}}{58} \approx 103\,\mathrm{nm}. \tag{3.2}$$

Hence, the bandwidth of the harmonics driven with 30 fs pulses at 800 nm wavelength in this free-focusing geometry is already significantly limiting the achievable resolution. It can be concluded that longer driving pulses would be needed in order to get a smaller bandwidth in the harmonics in view of approaching the Abbe limit.

The HHG-produced XUV light was subsequently used for most of the CDI and DIH experiments presented in this thesis together with the grating based imaging setup (Sect. 3.2.1).

[1] The spectrum is measured using a calibrated imaging spectrometer McPherson 248/310G equipped with a multi channel plate imaging unit.

[2] Perfect sampling of the diffraction pattern in terms of phase retrieval, i.e. $O = 2$, is assumed.

3.1.2 Fiber CPA Driven HHG Source

Fiber lasers have long been used to produce high average power laser light. Modern pulsed fiber lasers combined with a CPA scheme (FCPA) reach MHz level repetition rates and 200 μJ pulse energy, which results in an average power around 200 W [7], outracing modern Ti:Sa systems by at least a factor of 10 in average power. Having a typical pulse duration of 300–500 fs the peak power reaches the gigawatt level. Such pulses can thus be spectrally broadened in a hollow core fiber by self-phase modulation and subsequently compressed. The resulting ultrafast pulses are easily capable of generating high harmonics [8] at flux levels comparable to those reached with commercial Ti:Sa systems as presented in Sect. 3.1.1.

For the CDI experiments at high numerical aperture (Sect. 4.2) our group collaborated with the FIBER & WAVEGUIDE LASERS group (INSTITUTE OF APPLIED PHYSICS, FSU JENA), which pioneered the field of FCPA driven HHG sources. The FCPA laser system used, which is self-built by the group, is described in detail in [9]. The FCPA system delivers 660 fs pulses at $\lambda = 1{,}030$ nm central wavelength. After a subsequent hollow-core fiber compressor [7] 150 μJ pulses having 60 fs pulse duration are available. These pulses are focused ($f = 20$ cm) into a krypton jet for HHG. Typical intensities are in the range of $10^{14}\,\mathrm{W/cm^2}$ and are well suited to generate harmonics in krypton at 5 bar backing pressure. A chicane of two plane fused silica substrates is used to reduce the strong infrared pump before the water-cooled aluminum filters. Otherwise the high average power destroys the thin aluminum filters due to heat deposition. The HHG setup is depicted in Fig. 3.3a. The bandwidth of the 31st harmonic at 33.2 nm wavelength is approximately 0.15 nm. The relative bandwidth

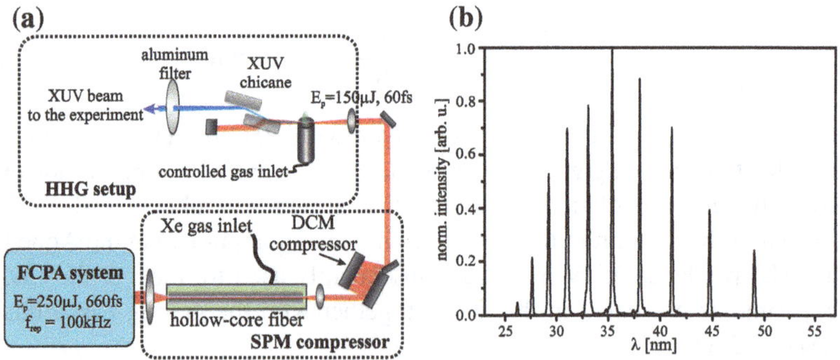

Fig. 3.3 The FCPA laser driven HHG source built by the FIBER & WAVEGUIDE LASERS group. **a** The output beam of an FCPA laser system is spectrally broadened by self-phase modulation in a xenon filled hollow-core fiber and subsequently compressed by means of dielectric chirped mirrors (DCM). In a krypton jet high harmonics are produced with a typical spectrum depicted in (**b**). The FCPA system and the HHG setup were built and optimized by the FIBER & WAVEGUIDE LASERS group. The presented spectrum and parameters are courtesy of the FIBER & WAVEGUIDE LASERS group (INSTITUTE OF APPLIED PHYSICS, FSU JENA)

$$\left.\frac{\Delta\lambda}{\lambda}\right|_{\lambda=33.2\,\mathrm{nm}} = \frac{0.15\,\mathrm{nm}}{33.2\,\mathrm{nm}} \approx \frac{1}{220} \tag{3.3}$$

is smaller compared to the Ti:Sa based setup (Sect. 3.1.1) due to the more than two times longer driving pulses. A typical spectrum is depicted in Fig. 3.3b. Following the consideration about a limitation of the achievable CDI resolution due to temporal coherence (cf. Eq. 3.2) allows a resolution estimate for the FCPA based system

$$\Delta r = \frac{Oa\Delta\lambda}{\lambda} = \frac{2 \times 3\,\mu\mathrm{m}}{220} \approx 27\,\mathrm{nm}. \tag{3.4}$$

Hence, one can conclude that the FCPA driven HHG source produces sufficiently narrow harmonic lines to achieve, in principle, sub-wavelength resolution. It is also worth mentioning that the excellent mode quality produced by the fiber based system is beneficial for CDI.

3.2 Table-Top CDI Schemes

3.2.1 *Grating Based CDI Setup*

After having discussed how ultrafast coherent XUV light is generated by means of a Ti:Sa laser (Sect. 3.1.1) or a fiber CPA laser (Sect. 3.1.2), a closer look on the actual CDI setups that were used for the experiments presented in this work is taken.

With synchrotrons usually crystals are used to monochromatize the radiation and pinholes are used to generate a source of limited size, yielding a defined radiation on the target [6]. With table-top HHG sources, however, one cannot use crystals which have a high absorption at typical HHG wavelengths. Moreover, the generated HHG radiation is divergent as the harmonics pick up the divergence of the generating infrared laser beam which is focused into a generation medium, i.e. argon and krypton for Ti:Sa HHG and FCPA HHG, respectively. Due to the limited flux that can be produced with table-top HHG sources one needs to take care of bringing as many of the photons onto the target as possible, since otherwise excessive exposure times would limit the usability.

A versatile setup that was built for this purpose [10] is depicted in Fig. 3.4a. For refocusing the harmonics onto the target a gold coated toroidal mirror ($f = 327.99\,\mathrm{mm}$, $\alpha_\mathrm{i} = 4.365°$)[3] is illuminated under grazing incidence, where gold has a reflectivity of over 87 % between 30 and 40 nm wavelength [11]. The mirror was placed in a $4f$-setup in order to image the HHG source one by one onto the target. Close behind the toroidal mirror a blazed gold grating with 158 lines/mm is placed, which is also hit under grazing incidence ($\alpha_\mathrm{i} \approx 4°$). The grating lines were oriented

[3] Throughout this thesis the angle of incidence is measured against the surface plane in order to be consistent with literature in the field of grazing-incidence small-angle X-ray scattering (GISAX).

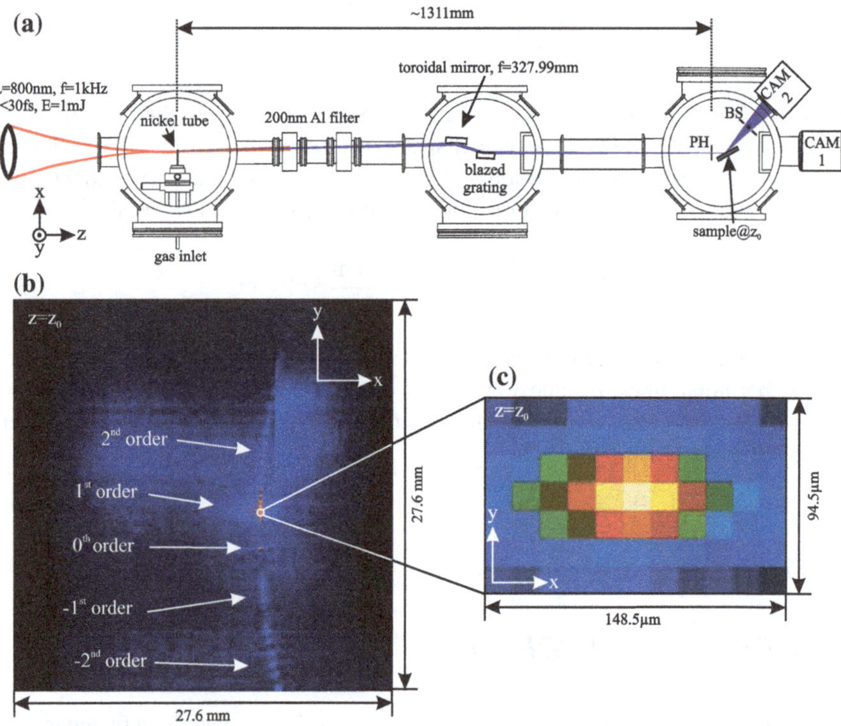

Fig. 3.4 The CDI setup (not to scale) using a toroidal mirror and grating to generate monochromatic light with high flux on the sample. **a** Overview of the setup as it is implemented in the lab. In the first chamber HHG radiation is produced as depicted in Fig. 3.2. An aluminum filter blocks all wavelengths longer than ~80 nm. In the central chamber a toroidal mirror placed in $2f$ distance from the source is illuminated under grazing incidence closely followed by a blazed grating. One harmonic focus is selected by means of a pinhole (PH, typical diameter ~100 µm) slightly before the focal plane. The sample is placed in the focus either in transmission geometry, i.e. using camera one (CAM1), or in reflection geometry, i.e. using camera two (CAM2). A beam stop (BS) can be inserted to suppress the strong central speckle. The overall CDI setup including the source is smaller than 2 m × 1 m. **b** A spatial measurement in the focal plane reveals the spatial distribution of the foci. One can see the different diffraction orders of the grating, while due to the blazed grating the first diffraction order is the strongest. The actual spectrum is comparable to the one shown in Fig. 3.2b. For the longer wavelengths of the first diffraction order there is an overlapping part with the second diffraction order, however, the strong harmonics around 36 nm wavelength are well separated. **c** A closer inspection of a single focus reveals an almost round shape with an approximate diameter of 50 µm. The strong pixelation is due to the limited size of one pixel on the used CCD. Such a focus can thus be selected by an, e.g. 100 µm, pinhole. Further spurious light originating from the non-perfect surface of the mirror and the grating (e.g. horizontal stripes in (**b**) to the left from each focus) can be efficiently suppressed

parallel to the optical table such that the spectral dispersion occurs in a perpendicular plane of the optical table. In the rear focal plane of the toroidal mirror the focused spectrum of the produced harmonics is observed (Fig. 3.4b). More exactly, the foci are arranged on the Rowland circle since this setup acts similar to a flat-field spectrometer.

However, the grating is arranged such that the lateral dispersion of the wavelengths in the rear focal plane in the first diffraction order is in the range of about 100 μm. Owing to the $4f$-setup the distance from the grating to the focal plane is ≈60 cm, hence the deviation from a flat plane is negligible. In order to maximize the photon flux for CDI experiments the grating was blazed such that the first diffraction order is preferred for the anticipated operation wavelengths, i.e. $\lambda \approx 30$–40 nm. The small dispersion introduced by the grating is just enough to separate the harmonics spatially, since the source size of the harmonics is typically in the range of a few tens of microns and the source is imaged one point at a time onto the back focal plane. For alignment purposes one can place an XUV sensitive CCD, i.e. an ANDOR IKON L having 4 megapixels with 13.5 μm ×13.5 μm pixel size on a roughly 1 × 1 in. chip for all experiments presented in this thesis, in the focal plane and optimize the harmonic yield and the performance of the toroidal mirror.

For CDI experiments one of the harmonics is selected. This is done by a 100 μm pinhole that is introduced just before the rear focal plane. Thus, behind the pinhole one gets a focus of monochromatic XUV radiation, which should have an almost flat phase in the focus and is of the size of the source, i.e. typically around 50 μm (FWHM) diameter (Fig. 3.4c). The pinhole also cleans up the beam profile at the cost that one usually gets small rings[4] on the recorded diffraction patterns. This point is discussed in the results chapter. An important issue is the wavelength spread inside the focus within a single harmonic due to the bandwidth of the harmonic. To get an estimate of the effect one can look at the values given above, where a lateral distance from one harmonic to the next around 36 nm wavelength of ≈100 μm was found. Linearizing the lateral distance from the 36 nm harmonic to the 39.34 nm harmonic (cf. Fig. 3.2) one gets roughly a value of $\approx 0.03\,\frac{\text{nm}}{\mu\text{m}}$ in the focus plane. Having measured the bandwidth $\Delta\lambda = 0.62$ nm, see Sect. 3.1.1, one can estimate that the lateral spread due to the bandwidth should be around ≈20 μm. Since the focal spot itself is larger than that, it is safe to expect the full bandwidth of a harmonic on the sample. However, for larger samples this could impose some problems, i.e. illuminating the sample with spatially chirped illumination. Practically, no limitation due to that effect was found in this work.

Behind the pinhole one can now place the sample either in a transmission or reflection geometry and capture the diffraction pattern afterwards. It should be noted that all of the previously described components are situated inside a vacuum chamber that is kept at a residual pressure of $<10^{-6}$ mbar. One of the main advantages of this setup using a toroidal mirror and a grating is that one can freely choose the wavelength out of the harmonics spectrum. Hence, one could for instance illuminate the same sample with different wavelengths one after another and, by doing so, benefit from spectral selective scattering [12, 13]. Moreover, one could use the zeroth diffraction order of the grating to have access to the full harmonic spectrum. The benefit would be a higher total flux, while at the same time a multi-wavelength code [14] is necessary to reconstruct the image. After all, it is useful to be able to select the harmonic that has the strongest flux, which can be variable depending on the laser's day-by-day

[4] These rings originate from diffraction of the outer parts of the beam at the round aperture.

performance. The relatively large focal spot, however, is limiting in the way that a lot of flux is lost for small samples, such as cells or nanoparticles, but at the same time it guarantees a homogeneous illumination of the sample.

The setup was used for the measurements in reflection geometry (Sect. 4.3), the breast cancer cell classification (Sect. 4.4) and the digital in-line holography experiments on non-periodic specimens (Sect. 4.1). For the latter only minimal modifications were made to the setup, which will be described in Sect. 3.3. Despite this CDI scheme being advantageous for many different applications, it has two major drawbacks. The first is the point by point imaging of the source onto the sample, because for very small samples having a size in the range of a micron one loses a lot of flux due to the relatively large focal spot.[5] The second problem is related to the round chambers used to build this setup (Fig. 3.4a). On the one hand the modular design gives a high degree of experimental freedom but on the other hand it limits the minimal distance from the sample to the CCD in reflection geometry and thus the numerical aperture. Practically, a resolution limit of about 1 μm is imposed by this restriction. Nevertheless, valuable results could be achieved with this setup, as will be discussed in the results chapter.

3.2.2 *Dielectric Mirror Based Setup*

From the limitations discussed in the previous subsection, i.e. an oversized focal spot for nanoscopic samples and limited numerical aperture, a new CDI chamber was designed in this work. It allows for high numerical aperture experiments in transmission and reflection geometry as well as for tight XUV foci. The new chamber was used in combination with the fiber CPA laser driven HHG source (Sect. 3.1.2) in order to benefit from the good spectral properties in the high NA experiments that aim to reach the Abbe limit.

The layout for the FCPA laser driven XUV source together with the high NA imaging chamber is depicted in Fig. 3.5a. Starting from the HHG source the beam first hits a chicane of two fused silica slides (angle of incidence $\sim 15°$) to dump some of the infrared light because otherwise the thin aluminum filters would be destroyed due to the heat deposition at roughly 15 W average power. The aluminum filter stage consists first of a water-cooled 200 nm filter which is followed by a 300 nm filter sitting on a sealed valve in order to have negligible ambient and infrared light passing into the imaging chamber. Two curved dielectric coated mirrors M1 ($f_{\mathrm{M1}} = 1{,}000$ mm) and M2 ($f_{\mathrm{M1}} = 500$ mm) are subsequently illuminated. M1 is placed in order to collimate the beam and M2 focuses the XUV light onto the sample. The spectral reflectivity of the two dielectric coated mirrors together is depicted in Fig. 3.5b (black solid curve) and allows the filtering of the 31st harmonic out of the HHG spectrum (blue dotted curve). It is worth noting that dielectric mirrors so far are the most popular

[5] Harder focusing of the HHG driving infrared laser would reduce the source size, but then the reduction of the interaction volume will substantially reduce the XUV flux.

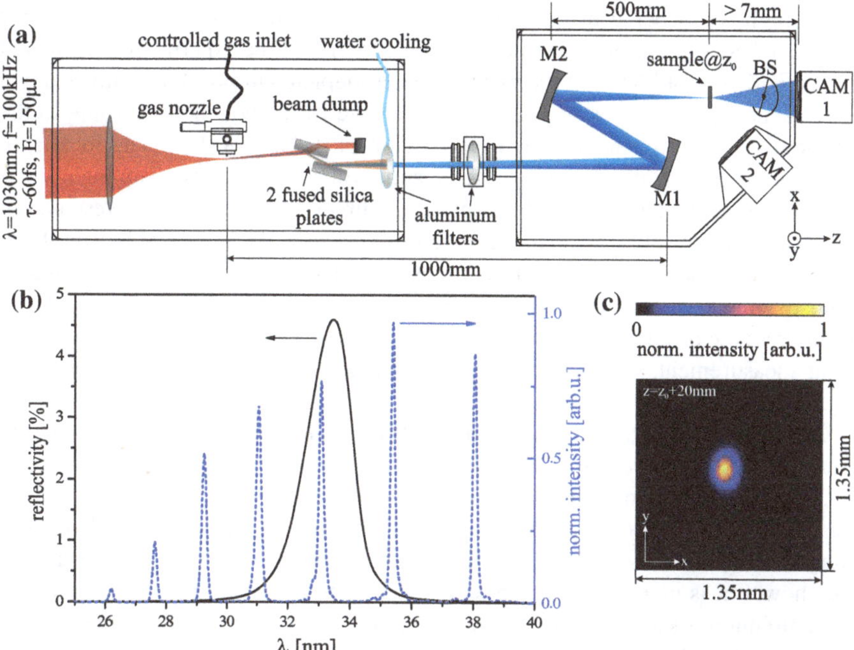

Fig. 3.5 The CDI setup (not to scale) using a set of dielectric mirrors for monochromatization. **a** Layout of the setup in combination with the FCPA laser driven HHG source, detailed in Sect. 3.1.2. After the Al filters the XUV radiation is collimated by M1 ($f = 1{,}000$ mm) and then focused down onto the sample by M2 ($f = 500$ mm). A beam stop (*BS*) can optionally be used to block the bright central speckle. Camera one (*CAM1*) can be used for high NA transmission geometry measurements. A specially designed mount can be used to bring a second camera (*CAM2*) close to the sample for reflection geometry measurements under $\alpha_i = 22.5°$. **b** Spectral properties of the dielectric mirror set (*black solid curve*). A peak reflectivity of ~4.6 % is achieved at 33.2 nm wavelength. This can be used to filter the 31st harmonic (*blue dotted curve*) of the HHG spectrum generated in krypton by the FCPA laser. **c** The shape of the beam measured 20 mm behind the focus features an almost symmetric Gaussian shape. The focal spot size was estimated to ~3 μm, see text for details. The spectrum and the reflectivity curve of the dielectric mirrors in (**b**) is courtesy of the Fiber & Waveguide Lasers group (Institute of Applied Physics, FSU Jena) and A. Guggenmos (LMU Munich) respectively

means of selecting a single wavelength out of a HHG spectrum for CDI, see e.g. in [15, 16]. Considering an angle of incidence on both mirrors of 83°, i.e. close to normal incidence, and p-polarization, one gets an overall reflectivity of approximately 4.6 %. The FWHM bandwidth of the mirror pair is ~1.9 eV which is sufficient to select a single harmonic and suppress the neighbor harmonics, considering a harmonics spacing of ~2.4 eV, at least by a factor of 20 compared to the selected harmonic. The z-fold setup of the two focusing mirrors (Fig. 3.5a) should be chosen as tight as possible, i.e. choosing the angle of incidence as close to normal incidence as possible, in order to reduce astigmatism which arises rapidly due to the short wavelength when being off-axis. A ray trace and an alternative mirror setup for minimized aberrations are presented in Appendix A. For the experiments conducted in this work a z-fold setup of the two mirrors, as depicted in Fig. 3.5a, with an angle of incidence $\geq 83°$ was employed. The achievable focal spot size according to the diffraction limit (Appendix A) is in the region of 2 μm (FWHM). The focal spot size was experimentally determined to be approximately 3 μm by scanning a small pinhole across the beam in the focal plane. A direct measurement, as presented in Fig. 3.4c, is not feasible due to the relatively large size of the CCD pixels. The shape of the beam about 20 mm behind the focus is depicted in Fig. 3.5c and features an almost symmetric Gaussian shape. A beam stop can be placed in front of the CCD to suppress the bright central speckle in order to increase the overall dynamic range of the measurement, see Sect. 3.2.3. A photographic view of the setup is depicted in Fig. A.2.

Compared to the grating based setup (cf. Fig. 3.4a) the dielectric mirror based setup has a fixed wavelength of operation. Since neighbor harmonics suppression is essential for imaging complex objects, a tight border condition is imposed on the HHG spectrum and the blue shifting of the harmonics can cause problems. For every driving laser a specific set of XUV mirrors must be inserted. The focal spot size, however, is more than a magnitude smaller and thus allows to bring the entire flux onto micron-sized samples. At the same time this requires a higher mechanical stability and is a limitation for extended samples. Using an even shorter focal length on M2, the focal spot size could be further reduced, which may allow for high-resolution ptychography in the near future. Another advantage worth mentioning is the high degree of fidelity for alignment in the setup, i.e. one can place the sample very close to the CCD, at the mechanical limits so as not to destroy the CCD, for a high numerical aperture and then translate M2 along the z-direction to place the focus exactly on the sample. That is beneficial compared to the grating based setup where the focal plane is fixed in the z-direction. A specialized camera mount together with a beveled side of the chamber allows for high NA measurements in reflection geometry. In summary, the new design provides the parameters expected to enhance high resolution CDI measurements together with the narrow-bandwidth high average power fiber CPA HHG source. This setup is employed for the high NA measurements that are presented in Sect. 4.2 of this work.

3.2.3 Enhancing the Dynamic Range of the Diffraction Pattern

For reconstructing high quality microscopic images from diffraction patterns the need for a good contrast of the experimental data, i.e. low noise and a high dynamic range, is obvious. However, it is well-known that the signal magnitudes in Fourier space strongly decrease for higher spatial frequencies.[6] In CDI this means that one typically gets a bright central speckle in the diffraction pattern, i.e. spatial frequency zero, and

[6] At least for the Fourier Transform of real-valued positive samples.

fringes with decreasing intensity towards the edge of the CCD.[7] For instance, the intensity of the Fourier transform of the grayscale image presented in Fig. 2.6c scales over roughly 7 orders of magnitude. However, commercially available detectors, such as the CCD that was used for the experiments, typically have a pixel depth of 16 bit corresponding to a maximum of 65,536 counts, i.e. roughly 5 orders of magnitude dynamic range. Unfortunately, as for the ANDOR IKON L, the noise caused by the readout electronics is typically between 300 and 1,000 counts, thus the effectively usable dynamic range is just around 3–4 orders of magnitude. Hence, the only way to experimentally sample a diffraction pattern over the full dynamic range using such detectors is by adding together single measurements taken with different exposure times. Another problem is that highly overexposed pixels on a CCD tend to overflow and spoil the surrounding pixels. Moreover, usually the linearity of the detectors becomes worse for count rates close to saturation.

In order to prevent these effects a beam stop is placed in front of the CCD to suppress the strong central part of the diffraction pattern for the longer exposure times. For shorter integration times the beam stop is removed in order to sample the central high intensity part of the diffraction pattern. The beam stop used in this work for the low NA reflection geometry measurements (Sect. 4.3) consisted of a 1 mm steel sphere supported by 15 μm tungsten wires affixed to a frame. For the high NA measurements (Sect. 4.2) a small sugar crystal (~100 μm diameter) fixed to a 5 μm tungsten wire was used. The reason for having differently sized beam stops for different numerical apertures is because of the relative size of the central speckle on the detector. It should be noted that the beam stop must be placed as close as possible to the detector in order to suppress diffraction on the beam stop itself. The procedure that is used for stitching together the diffraction patterns taken with different exposure times is as follows[8]:

1. Take the image with the longest exposure time and set all pixels having a value greater than 61,000 to zero.[9]
2. Subtract the background from the image. The background image, i.e. the laser turned off, is captured at all exposure times to account for permanent hotspots and the readout profile of the CCD.
3. Take the next shorter exposure time and apply steps 1–2 to it. Set all pixels under a certain threshold level, e.g. 100 counts, to zero.
4. Generate a mask that masks out zero intensity regions in the current overall image and zero intensity regions in the image to add.
5. Calculate a multiplication factor ν between the two images such that intensities of the unmasked region match best.

---

[7] If one considers a diffraction pattern centered on the detector.

[8] For the sake of simplicity the diffraction pattern will simply be denoted as *image*. The procedure starts with the image taken at the longest exposure time.

[9] The value 61,000 was found to be a good value for the high intensity cut-off of the ANDOR IKON L, for higher counts the linearity becomes poor.

6. Multiply the image with ν and average the unmasked regions with the current overall image while adding the multiplied image to regions where the current overall image has zero intensity.

Steps 3–6 are repeated until all captured images are accounted for. Using this procedure a diffraction pattern having a high dynamic range (HDR) is composed. Furthermore, the quality of the diffraction pattern is enhanced due to averaging parts of the diffraction patterns that were measured with different integration times. For the first image recorded without the beam stop an additional mask featuring the beam stop itself must be applied in step 4. The application of the procedure to experimental data is depicted in Fig. 3.6. From the experimentally determined multiplication factors from overlapping parts of the recorded patterns one can see that simply scaling by the exposure times would fail. The reason is that the flux of the HHG source is unstable and fluctuates from shot to shot. However, the dynamic range can be easily improved by about 4 orders of magnitude using this procedure and, thus one can assess information hidden for a single measurement due to the limited dynamic range of the CCD.

It is worth noting that currently efforts are undertaken to build CCDs specifically designed for X-ray diffraction experiments, i.e. having a large dynamic range in order to capture an appropriate diffraction pattern in one run [17].

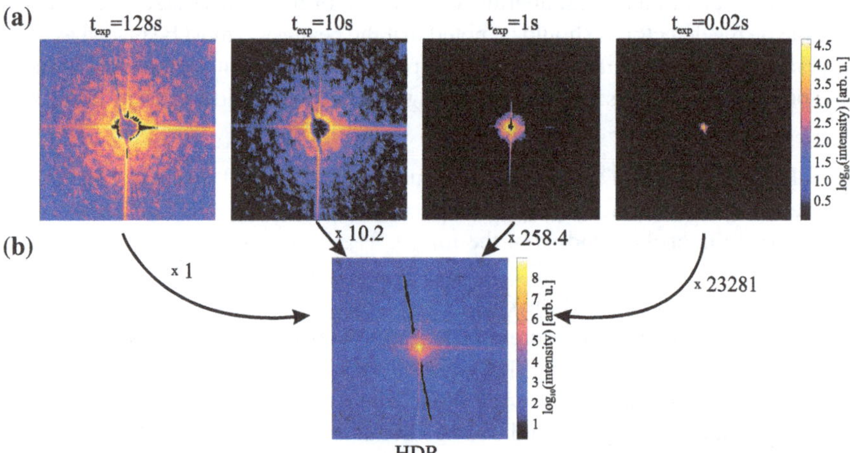

Fig. 3.6 Enhancing the dynamic range using a beam stop. **a** Raw measured diffraction patterns prepared according to steps 1 and 2 of the procedure, see text. A beam stop is used to suppress the overflow of the CCD from the central pixels for long exposure times ($t_{exp} = 128$ s and $t_{exp} = 10$ s). The central part of the diffraction pattern is then measured without the beam stop using shorter exposure times ($t_{exp} = 1$ s and $t_{exp} = 0.02$ s). Analyzing the scale, a dynamic range in the single measurement of roughly 4.5 orders of magnitude is evident. According to the procedure mentioned in the text, one can stitch all four together to get a high dynamic range (HDR) diffraction pattern, **b** spanning more than 8 orders of magnitude. Please note that the multiplication factors are determined by matching overlapping parts of different exposure times

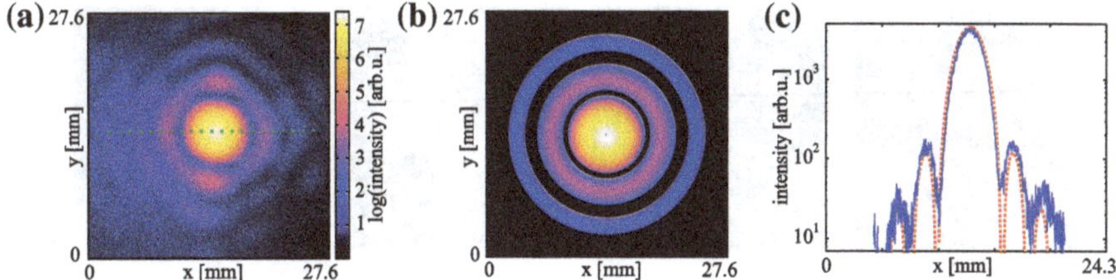

Fig. 3.7 Illumination wave formed by a 5 μm pinhole for DIH. **a** The recorded intensity 330 mm behind a 5 μm pinhole features the structure of an Airy pattern. **b** The spatial distribution of a best fit Airy pattern (maximum amplitude and noise-level cut-off adapted from the measured data) and the corresponding horizontal profile (*red dotted line* in **c**), profile position indicated by the *green dotted line* in (**a**) to the data presented in (**a**) can be used to determine the wavelength. In this case $\lambda = 39.0 \pm 0.5$ nm. The axes in (**a**) and (**b**) correspond to the full CCD chip

3.3 XUV Digital In-line Holography Realization

After having discussed the implementation of CDI into experiments, the implementation of digital in-line holography, which was theoretically introduced in the fundamental chapter Sect. 2.4, is the emphasis of this section. At this point it should be noted that the only experimental change compared to a CDI experiment is that a reference wave overlapping with the diffracted wave on the detector must be generated. In order to implement this, a small pinhole, e.g. 1 or 5 μm diameter, is introduced into the versatile grating based setup (Fig. 3.4a). In contrast to the description on the CDI setup (Sect. 3.2.1), one now places the small pinhole exactly in the focal plane to achieve a maximum flux transmitted through the pinhole. Obviously, a large amount of flux is lost considering a focal spot size of ~50 μm, but on the other hand a well-defined illumination wave is generated, see Fig. 3.7a. The additional modulation on the rings of the Airy pattern may have several origins, namely an imperfect aperture shape or the roughness of the inner surface. The central maximum, however, is smooth and clear and usable as the illumination wave for DIH. Fitting an Airy pattern (Fig. 3.7b, c) to the measured data can be used to determine the wavelength of the radiation quite accurately if the distance from the pinhole to the detector is known to ±1 mm, which is easily possible. In the case presented, $\lambda = 39.0 \pm 0.5$ nm is determined. The error was estimated from the accuracy of the fit. This method is beneficial in the grating based setup (Fig. 3.4a), where the wavelength can be chosen freely and where the electric driven translation stage, on which the pinhole was mounted, lacked feedback on the actual position.

According to the considerations given in Sect. 2.4, one places the detector at a distance from the pinhole such that the zeroth order of the Airy pattern produced by the pinhole uniformly illuminates[10] the central part of the detector. For the 5 and

[10] Uniform illumination means that the first minimum of the Airy pattern lies outside of the detector, i.e. a measurable intensity at all pixels. Obviously the distances would be much greater if one would aim for really homogeneous illumination. That would, however, reduce the usable flux too much for current table-top XUV systems.

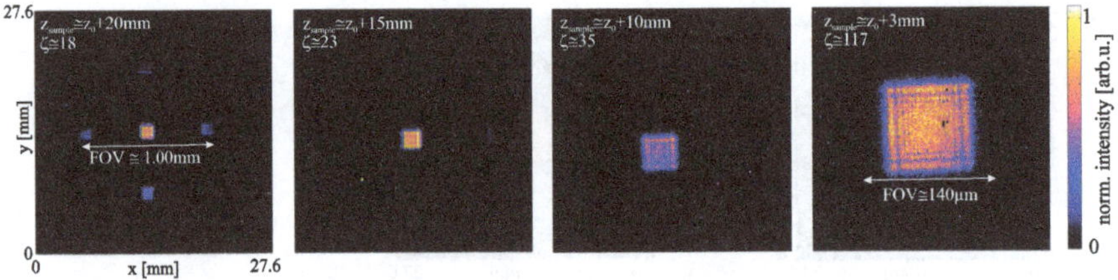

Fig. 3.8 Scanning the distance between pinhole and sample can be used to zoom-in and zoom-out, i.e. change the FOV and the magnification. The panels show digital in-line holograms of a blank Si_3N_4 aperture membrane (see Sect. 4.1) for the membrane being at different distances z_{sample} from a 1 μm pinhole which is located at z_0. The distance between the pinhole and the CCD was kept fixed at ~350 mm. The holograms are not properly resolved for reconstruction due to strong binning on the CCD, however, since the object consists of 100 μm × 100 μm square apertures which are 450 μm spaced apart one can at least estimate the field of view and the magnification ζ by relating the CCD pixels to the object. One can see the increasing magnification and decreasing FOV when the sample is brought closer to the pinhole while the distance of the CCD is fixed. The illumination was done using $\lambda = 39$ nm. Please note that the intensity scale is linear and the measures for z are roughly estimated due to missing position feedback of the translation stage. The x and y axes indicated on the *left panel* correspond to the full CCD chip and are the same for all panels

1 μm pinholes at 39 nm wavelength this is roughly 2,500 and 500 mm, respectively. The sample can either be placed in reflection geometry, such as is used for cancer cell classification (Sect. 4.4), or in transmission geometry, as it is used for imaging cell-like non-periodic specimens in this work (Sect. 4.1). The distance from the pinhole to the sample regulates the illuminated area that is projected onto the detector and thus regulates the FOV and magnification. Thus, one can comfortably change the magnification (Eq. 2.51) by changing this distance, this behavior is depicted in Fig. 3.8. Typical distances from the pinhole to the sample are of the order of a few millimeters.

3.4 Preparation of the Experimental Data

3.4.1 Intensity Normalization and Curvature Correction

In Sect. 3.2.3 the handling of experimental data in terms of background noise and limitations due to dynamic range of the detector was discussed. If available, a HDR diffraction pattern is always produced as the first step for all experiments presented in this work. Moreover, as discussed in Sect. 2.2, one can reconstruct an object from its continuous far-field diffraction pattern, which is related to the object plane by a Fourier transform. However, for a planar sample, this requires that the diffraction pattern, apart from proper sampling, is arranged such that each recorded pixel corresponds to a certain spatial frequency while the computation grid consists of pixels of equally spaced spatial frequencies. Likewise, for a three-dimensional sample, one needs a 3D grid with equally spaced spatial frequencies and the diffraction

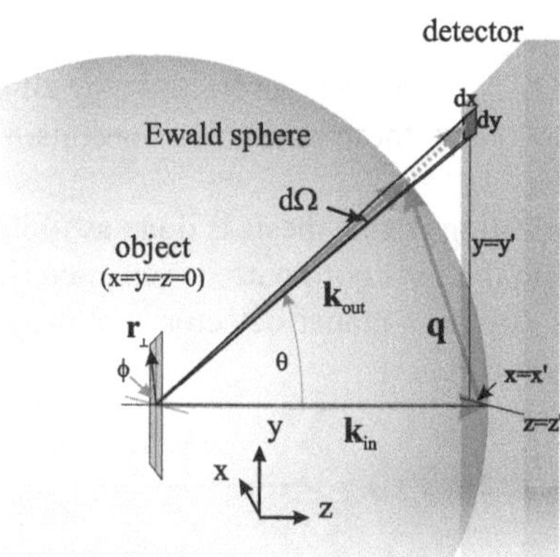

Fig. 3.9 Scattering in forward geometry at high numerical aperture. The scattered vector $\mathbf{k}_{\text{out}}$ ends on the Ewald sphere due to energy conservation (cf. Fig. 2.5). Using a planar detector the measured diffraction pattern must be remapped onto a sphere, this is called curvature correction. For a quantification (see text) spherical coordinates are projected onto the detector. θ denotes the angle between the incident $\mathbf{k}_{\text{in}}$ and scattered $\mathbf{k}_{\text{out}}$ wave vector. ϕ is the azimuthal vector in the object plane between the x-axis and the unit perpendicular projection vector of $\mathbf{k}_{\text{out}}$ which is denoted as $\mathbf{r}_\perp$ in the object plane. Furthermore, each pixel measures the photons emitted into a different solid angle $\mathrm{d}\Omega$. For instance a pixel of the size $\mathrm{d}x$ times $\mathrm{d}y$ at the edge of the detector (x', y', z') experiences a lower relative flux compared to one in the center of the CCD. The corresponding correction is called *intensity normalization*

pattern properly mapped onto the grid. As discussed in Sect. 2.2.2, all scattering vectors end on the Ewald sphere owing to energy conservation. Thus the diffraction pattern also lies on a sphere centered at the object. This issue is illustrated in Fig. 3.9. The diffraction patterns in this work are measured with a planar detector. Hence, as illustrated in Fig. 3.9, especially the fringes measured at the edges of the detector[11] have a larger distance from the center than they would have if they were sampled on an equally spaced spatial frequency grid on the Ewald sphere. Consequently, the measured diffraction pattern must be remapped onto the Ewald sphere, which is called *curvature correction* [18, 19]. Raines et al. showed that by remapping the diffraction intensities measured from a single view onto a sphere in a three-dimensional grid one can actually reconstruct a three dimensional object [20]. For two-dimensional samples, e.g. planar samples, it is, however, sufficient to do the inverse spherical projection and remain in a two-dimensional grid.[12] This greatly reduces the computation time for the phase retrieval compared to the 3D case. Another problem caused by the projection is that pixels having a different lateral distance from the center of the detector experience a different flux considering an object radiating homogeneously

[11] For the whole discussion in this section the zero deflection point, i.e. $\mathbf{k}_{\text{in}} = \mathbf{k}_{\text{out}}$ or $|\mathbf{q}| = 0$, is assumed in the center of the CCD.

[12] In Sect. 4.2 it will be discussed that for very high numerical apertures even hard apertures have a contribution from their inner surface and would thus need to be treated as a three-dimensional object.

in all directions. On the other hand, on a sphere every grid element corresponds to the same solid angle $\mathrm{d}\Omega$. This has to be corrected as well and it is called *intensity normalization* [20]. Obviously, these corrections are mainly necessary for a high numerical aperture.

The intensity normalization in this thesis is done as follows. For a pixel (e.g. the red square in Fig. 3.9 with the red coordinate set) characterized by its lateral size $\mathrm{d}x$ and $\mathrm{d}y$ at a position (x', y') on a planar detector at z' the solid angle $\mathrm{d}\Omega(x', y')$ is expressed in Cartesian coordinates

$$\mathrm{d}\Omega(x', y') = \frac{z'\mathrm{d}x\mathrm{d}y}{(x'^2 + y'^2 + z'^2)^{\frac{3}{2}}}. \tag{3.5}$$

Using this formula a matrix of solid angles $\mathrm{d}\Omega(\mathbf{x}, \mathbf{y})$ are computed for all detector pixels. Obviously, the solid angle per pixel decreases for pixels being further away from the center and would thus measure a lower intensity compared to the equally illuminated surface elements on the corresponding sphere (Fig. 3.10b). For the experimental data the resulting matrix of solid angles is normalized $\frac{\mathrm{d}\Omega(\mathbf{x},\mathbf{y})}{\mathrm{d}\Omega(0,0)} \rightarrow \mathrm{d}\Omega_{\mathrm{norm}}(\mathbf{x}, \mathbf{y})$, where $\mathrm{d}\Omega(0, 0)$ denotes the solid angle of the central pixel. At this point it should be noted that the rescaling according to the fact that the intensity falls by $1/r^2$ as used in [21] for intensity normalization is not sufficient and leads to a different result.

For the curvature correction one writes the momentum transfer vector $\mathbf{q}$ in terms[13] of the angle θ enclosed by $\mathbf{k}_{\mathrm{out}}$ and $\mathbf{k}_{\mathrm{in}}$, and the azimuthal angle ϕ being the angle between the x-axis and the projection of $\mathbf{k}_{\mathrm{out}}$ onto the sample plane (both angles are marked blue in Fig. 3.9)

$$\mathbf{q} = \mathbf{k}_{\mathrm{out}} - \mathbf{k}_{\mathrm{in}} = k\left[\sin\theta\cos\phi\hat{\mathbf{x}} + \sin\theta\sin\phi\hat{\mathbf{y}} + (\cos\theta - 1)\hat{\mathbf{z}}\right], \tag{3.6}$$

where $\hat{\mathbf{x}}$, $\hat{\mathbf{y}}$ and $\hat{\mathbf{z}}$ are the unit vectors in the corresponding direction. This is derived from expressing spherical coordinates in Cartesian coordinates. It can subsequently be rewritten into Cartesian coordinates (x, y, z) in the detector plane $z = z'$ [21]

$$\mathbf{q} = k\left[\frac{x}{\sqrt{x^2 + y^2 + z'^2}}\hat{\mathbf{x}} + \frac{y}{\sqrt{x^2 + y^2 + z'^2}}\hat{\mathbf{y}} + \left(\frac{z}{\sqrt{x^2 + y^2 + z'^2}} - 1\right)\hat{\mathbf{z}}\right]. \tag{3.7}$$

Identifying the two terms on the left as the q_x and q_y momentum transfer components respectively, one sees that they are not distributed uniformly over the detector. This is the result of projecting the Ewald sphere onto a planar detector. However, what is needed is a diffraction pattern that corresponds to a linear uniform sampling in q_x and q_y. In order to achieve this, q_x and q_y are computed for every pixel on

---

[13] Please note that the normalized wavevector $k_0 = 1/\lambda$ is used in order to obey the convention explained at the end of Sect. 2.2.2. The factor 2π would neither have an effect on the curvature correction nor the intensity normalization.

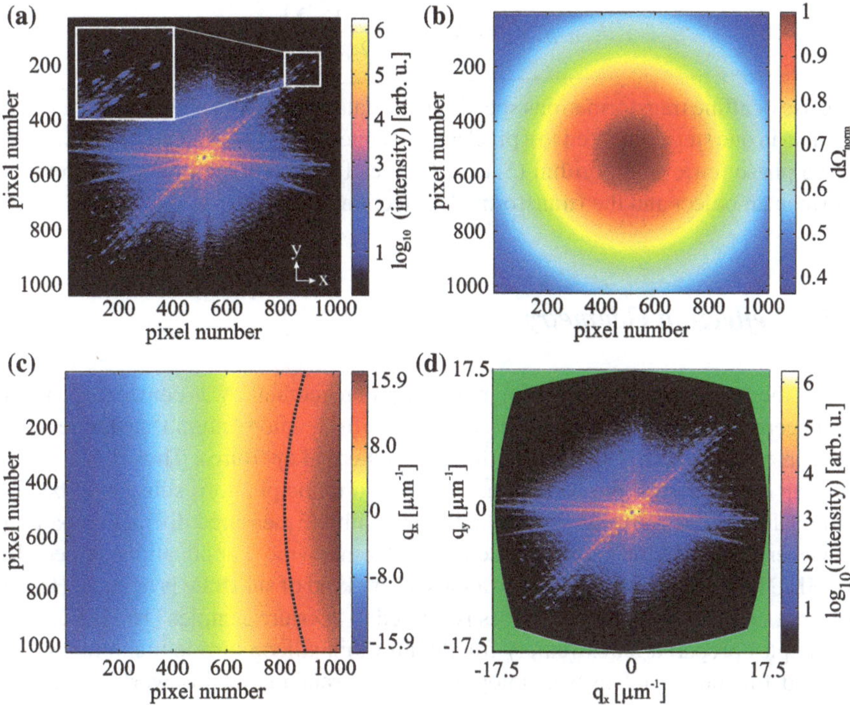

Fig. 3.10 Example for the data preparation in a high numerical aperture experiment. **a** A HDR raw diffraction pattern measured using the FCPA based setup (Fig. 3.5) operating at 33.2 nm wavelength. The distance from the sample to the detector was 20 mm, which corresponds to a NA of ≈0.57. The smeared fringes towards the edge of the detector due to the projection are clearly visible (*inset*). **b** The normalized solid angle per pixel. One can clearly see that the relative solid angle decreases radially from the center. Furthermore, one can see that the effect is far from negligible, because the pixels at the far edges of the detector cover less than half of the solid angle compared to the central pixel. **c** Depicts the mapping of the q_x component of the momentum transfer vector **q** for each pixel. The non-uniform mapping (one slice of constant q_x emphasized by the *dotted line*) is obvious, which is a result of the spherical projection. For the q_y component the mapping looks similar, just transposed. **d** After applying the intensity normalization and inverse spherical projection one can relate the measured data to the uniformly spaced **q**-space on the Ewald sphere. The *green regions* are zero density regions that arise in the inverse projection

the detector and, thus, this procedure results in a momentum transfer mapping for each component across the detector. Now an affine transformation matrix A can be computed using MATLAB, such that $A \cdot q_x(\mathbf{x}, \mathbf{y}) \rightarrow q'_{x\parallel}(\mathbf{x}, \mathbf{y})$, where $q'_{x\parallel}(\mathbf{x}, \mathbf{y})$ is now a linear momentum transfer mapping in q_x that is independent of the y position on the detector. In the same way one can compute an affine transformation matrix B for q_y. Having both matrices and the intensity correction map $\mathrm{d}\Omega_{\text{norm}}(\mathbf{x}, \mathbf{y})$ on hand one can then finally transform the measured diffraction pattern $I_m(\mathbf{x}, \mathbf{y})$ into one that samples linearly in $\mathbf{q}$ on the Ewald sphere with correct intensities

$$I_{\text{corr}}(\mathbf{x}, \mathbf{y}) = A \cdot B \cdot \frac{I_m(\mathbf{x}, \mathbf{y})}{\mathrm{d}\Omega_{\text{norm}}(\mathbf{x}, \mathbf{y})}. \tag{3.8}$$

The use of affine transformations speeds up the data preparation and allows to correct the data in one step compared to a pixel by pixel interpolation [21]. Feeding $I_{\text{corr}}(\mathbf{x}, \mathbf{y})$ into a phase retrieval algorithm (Sect. 2.3) gives full advantage of the FFT relation between an object and the (transformed) detection plane.

3.4.2 Reflection Geometry

In this thesis so far only diffraction in transmission geometry was considered, which is the geometry of choice since the components of the momentum transfer vector $\mathbf{q}$ scale as symmetric and even linear for low numerical apertures. Thus if NA $\lesssim 0.1$ one can even leave all the corrections introduced in the previous subsection behind and feed the phase retrieval algorithm directly with the measured diffraction pattern. However, the reflection geometry seems to offer many more applications, especially for HHG based CDI systems, where almost every kind of substrate is opaque. Hence, transmission CDI with HHG sources is limited to aperture samples [16, 18, 22, 23] or samples prepared on extremely fragile ultra-thin silicon nitride membranes [24] as used for the digital in-line holography measurements in this work (Sect. 4.1). In reflection geometry however, one can use almost any kind of sample as long as it partially reflects XUV light, e.g. an XUV reflective sample on an absorbing substrate, an absorbing sample on a reflective substrate or a reflective sample on a reflective substrate. The preparation is straightforward, e.g. the samples can be pipetted onto the substrate, such as the breast cancer cells in this work (Sect. 4.4). Also imaging of nanostructures that are fabricated by electron lithography (Sect. 4.3) becomes possible. Thus, the reflection geometry offers more applications, especially for experiments where high temporal resolution is needed, e.g. for imaging plasmon waves in optical nanostructures.

However, the drawback of the reflection geometry is that an additional momentum transfer in one axis arises from the reflection. Thus independent from the NA one will have to make corrections to the measured data, while the exact correction parameters are sometimes hard to determine in an experiment. Furthermore, usually sample mounts and camera mounts cannot be brought together sufficiently close to achieve high numerical apertures compared to a sandwich-like setup in transmission geometry. Another issue in reflection geometry is that the surface roughness of the sample and the substrate contribute to the diffraction pattern and will in some cases even dominate the measured signal. The surface roughness must be kept well below $\lambda/10$ over an area larger then the focal spot size, which can be challenging for XUV wavelengths and even more for soft X-rays. Maybe these are the reasons why it took until 2012 when two groups published first CDI experiments in reflection geometry almost at the same time [21, 25], followed by a publication for low NA table-top CDI in reflection geometry [26].

If one considers a beam incident on the sample under an angle of incidence α_i and observation in reflection geometry (Fig. 3.11), one has to use an appropriate coordinate transformation using a rotation matrix around the y-axis [21]. Moreover, a transformation from the lab system to the coordinate system of the sample is useful, because unlike in the case presented before, the detector plane is tilted to the sample plane. Thus, one now analyzes the horizontal exit angle α_f which is in plane with the angle of incidence α_i and the vertical exit angle θ_f with respect to the sample plane and expresses the momentum transfer vector $\mathbf{q}$ by these. In contrast to the previous section it is now considered that the CCD is centered in the specular reflection point, i.e. when $\alpha_f = \alpha_i$ and $\theta_f = 0$. The components of the momentum transfer in reflection geometry then become [25, 27]

$$q_x = k\left[\cos\alpha_f \cos\theta_f - \cos\alpha_i\right] \tag{3.9}$$

$$q_y = k\left[\cos\alpha_f \sin\theta_f\right] \tag{3.10}$$

$$q_z = k\left[\sin\alpha_i + \sin\alpha_f\right]. \tag{3.11}$$

This representation of the momentum transfer vector components is convenient, because typically in a reflection geometry experiment the angle between the incident beam and the detector is fixed by the chamber geometry and one simply places the specular reflection in the center of the detector. The notation used here, in contrast to the previous notation for the transmission geometry, is also consistent with other fields that take advantage of reflective scattering such as neutron scattering [28] and grazing-incidence small-angle X-ray scattering (GISAX) [27]. Inspecting Eqs. 3.9 and 3.10 one finds zero momentum transfer in q_x and q_y at the point of specular reflection. One should, however, keep in mind that there is always a constant momentum transfer in the lab system due to the reflection. Moreover, q_z is mapped on the detector in the same direction as q_x and thus one gets information from the height profile superimposed with the information from one lateral dimension of the sample. Especially for tiny angles of incidence and samples with a distinctive height profile, this must be corrected [25]. On the other hand, one could use the information to map the measured diffraction pattern onto a 3D grid and perform a full 3D reconstruction in the same manner as illustrated in [20]. Keeping all this in mind, θ_f and α_f are computed for every detector pixel and subsequently used to compute the maps of the components q_x, q_y and q_z in order to correct the experimental data.

The reflection geometry experiments reported in this work were done at an angle of incidence $\alpha_i = 22.5°$ using the grating based setup (Fig. 3.4). Due to the available chamber construction the minimum distance from the sample to the detector was limited to $L_{\text{obj-CCD}} = 430\,\text{mm}$, which results in an numerical aperture NA ≈ 0.032. The corresponding mapping of the components of the momentum transfer vector are depicted in Fig. 3.12. One can see the non-uniform and asymmetric behavior for the q_x component as expected from Eq. 3.9. The equi-momentum transfer slice $q_x = 0$ is marked by the dotted purple line in Fig. 3.12a. In the presented case this effect is minimal and will thus be neglected for the analysis of the experimental data, however, for higher NAs this becomes severe and must be corrected (Fig. 3.12d–f).

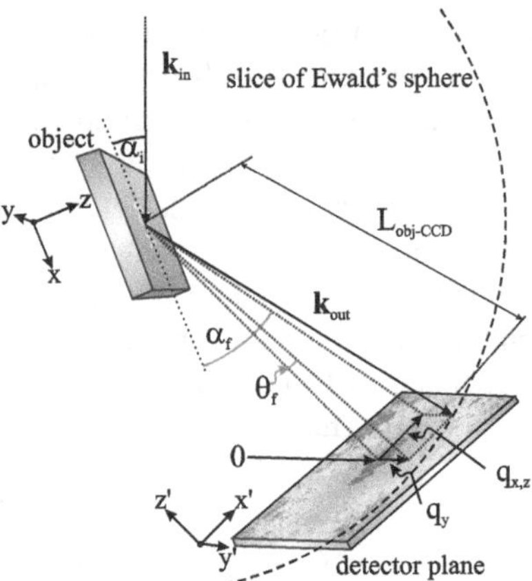

Fig. 3.11 Definition of the reflection geometry for XUV CDI. Analogously to the transmission geometry (Fig. 2.5) one is interested in the mapping of components of the momentum transfer vector $\mathbf{q} = (q_x, q_y, q_z)$ onto the Ewald sphere, which equals the detector for a low NA as typical in this work. The object plane and detector plane are rotated towards each other (see the two coordinate systems). Thus it makes sense to calculate the momentum transfer in the sample frame in terms of the horizontal α_f and vertical θ_f exit angles. By knowing the angle of incidence α_i and that the detector is placed perpendicular on the intersection with the Ewald sphere and centered at the point of specular reflection (marked by 0) one can calculate the momentum transfer components, see text. The distance from the center of the object to the CCD is denoted as $L_{\text{obj-CCD}}$

The mapping of the q_y component (Fig. 3.12b) is in contrast to that uniformity and features an approximately 2.5 times higher magnitude compared to the q_x component. This illustrates the effect that the detector observes the object projected under an angle of 22.5°, hence it appears $1/\sin(22.5°) \approx 2.61$ times reduced in the horizontal direction. The q_y mapping is otherwise desirably linear at this low NA and needs no additional correction. For a high NA one would have similar effects as in transmission geometry and would need to correct for the curvature additionally. The q_z component mapping (Fig. 3.12c) reveals the problem discussed before, i.e. the q_z component is linearly projected onto the horizontal detector axis while the mean momentum transfer in this component $q_z \approx 20\,\mu\text{m}^{-1}$ corresponds to a constant momentum transfer added by the reflection. The observable magnitude for q_z from the center to the right detector edge is approximately $\Delta q_z \approx 0.8\,\mu\text{m}^{-1}$ and thus, using Eq. 2.31, the achievable resolution in the height profile of the object can be determined to $\Delta r_z \approx 0.63\,\mu\text{m}$. Hence, this superimposed mapping of q_x and q_z causes only trouble for objects significantly higher than $\sim 0.5\,\mu\text{m}$, otherwise no additional fringes arise from the light reflected from the highest level and the lowest level of the object. This condition is fulfilled for the measurements presented in this work (Sect. 4.3).[14] Hence, in the remainder of this thesis the q_z component will not be considered (Eq. 3.11).

[14] For the biological specimen classification (Sect. 4.4) this condition does not hold true, because carbon has a reflectivity comparable to that of gold for the wavelengths and angle of incidence used here [11]. However, no reconstruction is anticipated in that part of the work and thus the scaling in $\mathbf{q}$ can be disregarded there.

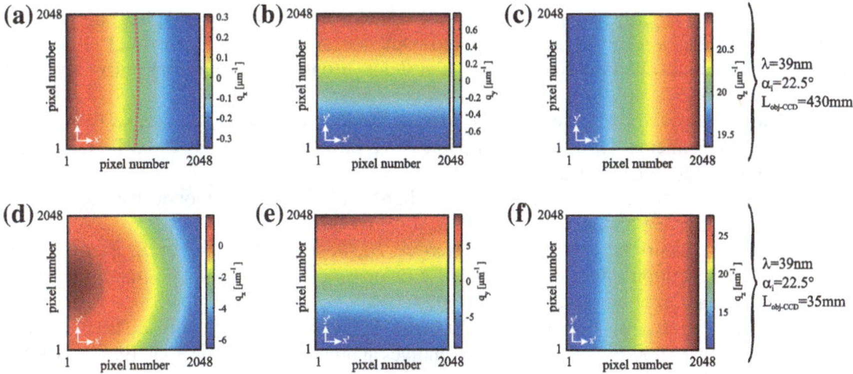

Fig. 3.12 Mapping of the components of the momentum transfer vector for the reflection geometry experiments presented in this work (angle of incidence $\alpha_i = 22.5°$, $\lambda = 39\,\text{nm}$ wavelength, NA ≈ 0.032) in *panels* (**a–c**) and for a theoretical high NA experiment in *panels* (**d–f**). **a** The mapping of the q_x component is slightly non-uniform in the vertical direction (i.e. direction of the q_y component) and asymmetric in horizontal direction. The *purple dotted line* guides the eye on $q_x = 0$. **b** The mapping of q_y instead is linear, uniform and symmetric. **c** Momentum transfer map for the q_z component. It is linear and uniform and centers around a constant momentum transfer $q_z \approx 20\,\mu\text{m}^{-1}$ given by the reflection itself. In *panels* (**d–f**) the corresponding mappings for a high NA experiment are depicted (NA ≈ 0.37), while all other parameters are kept constant. Note that the left edge of the detector (pixel number 1) in (**d**) corresponds to scattering along the object plane (i.e. $\alpha_f \approx 0°$). Furthermore, the scaling in q_x becomes extremely asymmetric at such a high NA. Please note that mapping corresponds to a detector tied to the Ewald sphere and that curvature correction effects (see section before) are not accounted for. Hence, the *panels* (**d–f**) just illustrate the idea

In the panels d–f in Fig. 3.12 theoretical mappings for a high NA ($\approx$0.37) experiment in reflection geometry are presented. The q_x component mapping, which geometrically extends towards the object plane, i.e. $\alpha_f \approx 0°$, is severely distorted and asymmetric. Likely, this asymmetry will effectively limit the achievable resolution of this geometry in the object's x-axis, because in order to linearize the q_x component one could just use a narrow band of data between $\pm 1.8\,\mu\text{m}^{-1}$ resulting in a resolution around $\Delta r_x \approx 280\,\text{nm}$ in this specific case. Please note that the panels in Fig. 3.12 are corresponding to a detector tied to the Ewald sphere, which is a good estimate for the panels a–c, but obviously not for d–f. Hence, an additional curvature correction, as discussed in the previous section, would be necessary to account for the planar detector at high NAs.

References

1. Diels, J.C., Rudolph, W.: Ultrashort Laser Pulse Phenomena: Fundamentals, Techniques, and Applications on a Femtosecond Time Scale. Academic Press, London (2006)
2. Spielmann, C., Xu, L., Krausz, F.: Measurement of interferometric autocorrelations: comment. Appl. Opt. **36**(12), 2523–2525 (1997)

3. Trebino, R., Kane, D.J.: Using phase retrieval to measure the intensity and phase of ultrashort pulses—frequency-resolved optical gating. J. Opt. Soc. Am. A **10**(5), 1101–1111 (1993)
4. Brabec, T., Krausz, F.: Intense few-cycle laser fields: Frontiers of nonlinear optics. Rev. Mod. Phys. **72**(2), 545–591 (2000)
5. Hergott, J.F., Kovacev, M., Merdji, H., Hubert, C., Mairesse, Y., Jean, E., Breger, P., Agostini, P., Carre, B., Salieres, P.: Extreme-ultraviolet high-order harmonic pulses in the microjoule range. Phys. Rev. A **66**(2), 021801 (2002)
6. Ishikawa, T., Tamasaku, K., Yabashi, M.: High-resolution X-ray monochromators. Nucl. Instrum. Meth. A **547**(1), 42–49 (2005)
7. Rothhardt, J., Haedrich, S., Carstens, H., Herrick, N., Demmler, S., Limpert, J., Tuennermann, A.: 1 MHz repetition rate hollow fiber pulse compression to sub-100-fs duration at 100 W average power. Opt. Lett. **36**(23), 4605–4607 (2011)
8. Haedrich, S., Krebs, M., Rothhardt, J., Carstens, H., Demmler, S., Limpert, J., Tuennermann, A.: Generation of microW level plateau harmonics at high repetition rate. Opt. Express **19**(20), 19374–19383 (2011)
9. Eidam, T., Rothhardt, J., Stutzki, F., Jansen, F., Haedrich, S., Carstens, H., Jauregui, C., Limpert, J., Tuennermann, A.: Fiber chirped-pulse amplification system emitting 3.8 GW peak power. Opt. Express **19**(1), 255–260 (2011)
10. Zuerch, M., Spielmann, C.: Verfahren und Vorrichtung zur Erzeugung einer schmalbandigen, kurzwelligen, kohaerenten Laserstrahlung, insbesondere fuer XUV-Mikroskopie, DE102012022961.5, 21 Nov 2012 (2012)
11. Henke, B.L., Gullikson, E.M., Davis, J.C: X-ray interactions: photoabsorption, scattering, transmission, and reflection at E=50-30000 eV, Z=1-92. At. Data Nucl Data. Tables **54** (2), 181–342 (1993)
12. Eisebitt, S., Luning, J., Schlotter, W.F., Lorgen, M., Hellwig, O., Eberhardt, W., Stohr, J.: Lensless imaging of magnetic nanostructures by X-ray spectro-holography. Nature **432**(7019), 885–888 (2004)
13. Weise, F., Neumark, D.M., Leone, S.R., Gessner, O.: Differential near-edge coherent diffractive imaging using a femtosecond high-harmonic XUV light source. Opt. Express **20**(24), 26167–26175 (2012)
14. Chen, B., Dilanian, R.A., Teichmann, S., Abbey, B., Peele, A.G., Williams, G.J., Hannaford, P., Van Dao, L., Quiney, H.M., Nugent, K.A.: Multiple wavelength diffractive imaging. Phys. Rev. A **79**(2), 023809 (2009)
15. Sandberg, R.L., Paul, A., Raymondson, D.A., Haedrich, S., Gaudiosi, D.M., Holtsnider, J., Tobey, R.I., Cohen, O., Murnane, M.M., Kapteyn, H.C., Song, C.G., Miao, J.W., Liu, Y.W., Salmassi, F.: Lensless diffractive imaging using tabletop coherent high-harmonic soft-X-ray beams. Phys. Rev. Lett. **99**(9), 098103 (2007)
16. Zhang, B., Seaberg, M.D., Adams, D.E., Gardner, D.F., Shanblatt, E.R., Shaw, J.M., Chao, W., Gullikson, E.M., Salmassi, F., Kapteyn, H.C., Murnane, M.M.: Full field tabletop EUV coherent diffractive imaging in a transmission geometry. Opt. Express **21**(19), 21970–21980 (2013)
17. Potdevin, G., Trunk, U., Graafsma, H.: Performance simulation of a detector for 4th generation photon sources: the AGIPD. Nucl. Instrum. Method A **607**(1), 51–54 (2009)
18. Sandberg, R.L., Song, C., Wachulak, P.W., Raymondson, D.A., Paul, A., Amirbekian, B., Lee, E., Sakdinawat, A.E., La, O., Vorakiat, C., Marconi, M.C., Menoni, C.S., Murnane, M.M., Rocca, J.J., Kapteyn, H.C., Miao, J.: High numerical aperture tabletop soft X-ray diffraction microscopy with 70-nm resolution. Proc. Natl. Acad. Sci. USA **105**(1), 24–27 (2008)
19. Sandberg, R.L.: Closing the gap to the diffraction limit: near wavelength limited tabletop soft X-ray coherent diffractive imaging. Ph.D. thesis, University of Colorado (2009)
20. Raines, K.S., Salha, S., Sandberg, R.L., Jiang, H., Rodriguez, J.A., Fahimian, B.P., Kapteyn, H.C., Du, J., Miao, J.: Three-dimensional structure determination from a single view. Nature **463**(7278), 214–217 (2010)
21. Gardner, D.F., Zhang, B., Seaberg, M.D., Martin, L.S., Adams, D.E., Salmassi, F., Gullikson, E., Kapteyn, H., Murnane, M.: High numerical aperture reflection mode coherent diffraction microscopy using off-axis apertured illumination. Opt. Express **20**(17), 19050–19059 (2012)

22. Seaberg, M.D., Adams, D.E., Townsend, E.L., Raymondson, D.A., Schlotter, W.F., Liu, Y.W., Menoni, C.S., Rong, L., Chen, C.C., Miao, J.W., Kapteyn, H.C., Murnane, M.M.: Ultrahigh 22 nm resolution coherent diffractive imaging using a desktop 13 nm high harmonic source. Opt. Express **19**(23), 22470–22479 (2011)
23. Ravasio, A., Gauthier, D., Maia, F.R., Billon, M., Caumes, J.P., Garzella, D., Geleoc, M., Gobert, O., Hergott, J.F., Pena, A.M., Perez, H., Carre, B., Bourhis, E., Gierak, J., Madouri, A., Mailly, D., Schiedt, B., Fajardo, M., Gautier, J., Zeitoun, P., Bucksbaum, P.H., Hajdu, J., Merdji, H.: Single-shot diffractive imaging with a table-top femtosecond soft X-ray laser-harmonics source. Phys. Rev. Lett. **103**(2), 028104 (2009)
24. Seibert, M.M., Boutet, S., Svenda, M., Ekeberg, T., Maia, F.R.N.C., Bogan, M.J., Timneanu, N., Barty, A., Hau-Riege, S., Caleman, C., Frank, M., Benner, H., Lee, J.Y., Marchesini, S., Shaevitz, J.W., Fletcher, D.A., Bajt, S., Andersson, I., Chapman, H.N., Hajdu, J.: Femtosecond diffractive imaging of biological cells. J. Phys. B-at Mol. Opt. **43**(19), 194015 (2010)
25. Sun, T., Jiang, Z., Strzalka, J., Ocola, L., Wang, J.: Three-dimensional coherent X-ray surface scattering imaging near total external reflection. Nat. Photon. **6**(9), 588–592 (2012)
26. Zuerch, M., Kern, C., Spielmann, C.: XUV coherent diffraction imaging in reflection geometry with low numerical aperture. Opt. Express **21**(18), 21131–21147 (2013)
27. Smilgies, D.M.: High-resolution grazing-incidence scattering using a combination of analyzer crystal and linear detector. Rev. Sci. Instrum. **74**(9), 4041–4047 (2003)
28. Paul, A.: Grazing incidence polarized neutron scattering in reflection geometry from nanolayered spintronic systems. Pramana-J. Phys. **78**(1), 1–58 (2012)

Chapter 4
Lensless Imaging Results

In the previous chapters the fundamentals for coherent diffraction imaging and digital in-line holography were described. Moreover, the experimental techniques and the treatment of the experimental data were introduced. In this chapter the focus lies on the experimental results that were obtained in this work. The digital in-line holography experiments were carried out with the grating based setup. Special cell-like samples were prepared on thin silicon nitride membranes. Besides the preparation methods of these samples, the results for the DIH experiments are presented in Sect. 4.1. The novel dielectric mirror based setup (Sect. 3.2.2 and Appendix A) was used together with the FCPA HHG source (Sect. 3.1.2) for transmission CDI experiments at a high numerical aperture targeting for a resolution in the range of the wavelength or even below. These high NA experiments can be found in Sect. 4.2. It was explained before that the reflection geometry offers unique possibilities in CDI, because samples can be prepared more easily and one gets access to the surface of objects that are opaque and can thus not be imaged in transmission geometry. The experiments were carried out by imaging micron scaled objects in reflection geometry (Sect. 3.4.2) using the grating based setup (Sect. 3.2.1). The results for imaging a non-periodic specimen in reflection geometry are presented in Sect. 4.3. Finally, the same setup, but with a smaller pinhole in order to go into a mixed CDI/DIH regime, was used to measure diffraction patterns of human breast cancer cells. It was found that one can use these diffraction patterns to classify the cells and thus provide a tool that may allow fast cell type discrimination in a clinical environment. The results on the human breast cancer cell experiments are presented in Sect. 4.4.

4.1 Digital In-line Holography of Cell-Like Non-periodic Specimens

One of the main limitations of CDI discussed in the fundamentals is that samples must be isolated, i.e. surrounded by a zero density region, to apply the support constraint. Despite this limitation remarkable results were achieved using synchrotrons

M.W. Zürch, *High-Resolution Extreme Ultraviolet Microscopy*, Springer Theses, DOI 10.1007/978-3-319-12388-2_4

[1] and free-electron lasers [2], proving the importance of applying these techniques to biological and medical research. However, for a daily clinical application, the need for synchrotron radiation and the need for isolated samples are criteria for exclusion. A way to overcome the isolation constraint, as discussed in Sect. 2.5, is ptychography [3], which, however, needs scanning of the sample which may be feasible for imaging a single sample. Though, for daily application in a medical environment, where a high number of samples must be investigated in a short time, this approach seems not to be practical. Moreover, one would again struggle with the limited dynamic range of the detector and thus need to use beam stops and different exposure times to measure a diffraction pattern.

Instead, the approach presented here is digital in-line holography, which was introduced in Sect. 2.4. By illuminating the sample with a curved reference wave one gets a hologram in the far-field that features typically just small intensity changes across the full detection plane. Moreover, data collection is not critical, since hot spots and spurious pixels may effect the result but not the convergence of the reconstruction. And finally, the reconstructions can be done in real-time offering high fidelity and throughput in a clinical environment. At this point one has to mention that the radiation having wavelengths of several 10 nm is highly absorbed by all biological matter, because it is far away from the water window.[1] Hence, in order to investigate on the fundamental principles and the image reconstruction, samples that are not isolated and exhibit sizes and structures of biological specimens were prepared. However, measures had to be taken, such that a part of the samples remains transparent. The results for preparing such samples are presented in the following section. In Sect. 4.1.2 subsequently the obtained results with those samples are presented.

4.1.1 Cell-Like Sample Preparation for DIH XUV Experiments

For the DIH experiments in this work, the transmission geometry using a wavelength between 30 and 40 nm was chosen, hence a transparent substrate at these wavelengths is needed. Silicon nitride (Si_3N_4) is a material that has successfully been used at these wavelengths before [4]. Thin silicon nitride membranes are commercially available for transmission electron microscopy. The membranes are typically fabricated by depositing Si_3N_4 on a bulk silicon substrate[2] followed by etching down bevels into the silicon wafer, such that a small window, consisting of the Si_3N_4 film, remains. The substrates used in this work are from Ted Pella Inc. with a Si_3N_4 membrane thickness of 15 nm. The Si_3N_4 windows on each substrate are arranged in a 3 by 3 array with each window being 100 by 100 μm wide (Fig. 4.1). The transparency of 15 nm Si_3N_4 at 38 nm wavelength is approximately 44 % [5] and thus forms a well suited substrate for the experiments done in this work. Surprisingly, the 15 nm thick

[1] The water window is a range of wavelengths between the K-edge of oxygen at 2.34 nm and the K-edge of carbon at 4.4 nm, where biological materials become transparent.

[2] A 3 mm diameter silicon disk with 200 μm thickness in this case.

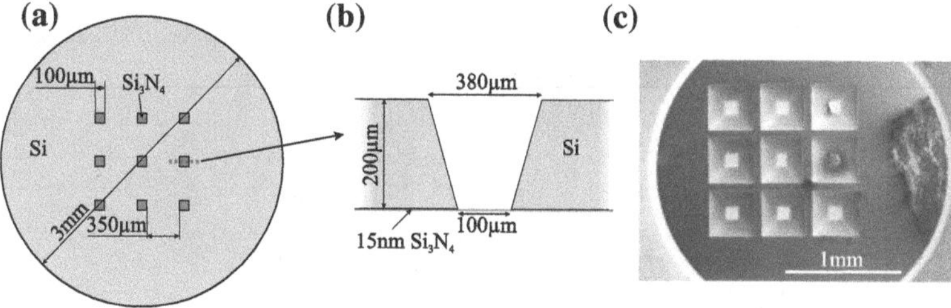

Fig. 4.1 Characteristics of the used Si_3N_4 substrates. **a** *Top view* of the substrate. **b** *Side view* cut (not to scale) at the position of the *dotted green line* in (**a**). At the places where the *square shaped bevels* were cut into the silicon wafer only a 15 nm thin Si_3N_4 membrane is left. **c** A scanning electron microscope (SEM) image of a substrate mounted onto an aluminum mount ready to be mounted in the target chamber (Fig. A.2)

membranes are relatively stable and even withstand large liquid drops deposited by a pipette.

The nanoparticles that were employed to mimic cells are uniform polystyrene latex beads from MAGSPHERE INC. These are commercially available in a suspension with the size of the beads ranging from several nanometers up to several microns. Roughly classifying objects of interest in medicine, one can look at cells (typical sizes several microns), bacteria (typical size around 1 μm) and viruses (typical sizes of several 10 nm up to several 100 nm) [6]. Hence, in order to prepare the samples mimicking bacteria and cells beads with 900 nm and 5 μm diameter respectively were used. Smaller structures mimicking viruses were not considered, because the resolution achievable with DIH at the stage of this thesis would be insufficient to resolve them. The challenging task was to bring the beads from the suspension onto the Si_3N_4 membrane, without destroying the membrane and without leaving dirt from the solvent on the membrane. Several methods were tested [7]. It was found that the best way was to print them onto the substrate using the printhead of an inkjet printer. In order to do this, a solution of the beads was thinned with distilled water and filled into the printhead which had all mechanical filters removed. Applying a short voltage trigger ($t \approx 2\,\mu s$) fires a pre-loaded solution drop towards the membrane. Since this process is initiated by instantly heating a small resistor, it heats up the solution as well and most of the water evaporates on its way onto the sample through ambient air. Hence, one can deposit the beads almost without solvent remains onto the substrate. Pipetting, using high precision pipettes, failed, because the minimal applicable volume is in the range of a few 100 nl, which already entirely floods a bevel. This results in all nanoparticles being dragged off the membrane when the water, which is dragged up the bevel walls due to surface tension, evaporates. Pipetting on the flat substrate side spills the droplet over the whole wafer and an exact placement of the beads is not possible. Moreover, solvent remains usually spoil the membrane in terms of XUV transparency (Fig. 4.2a). Printing, however, worked well and by adjusting the amplitude and temporal length of the voltage trigger one could adjust the amount of solution fired onto the sample and, hence, control the amount

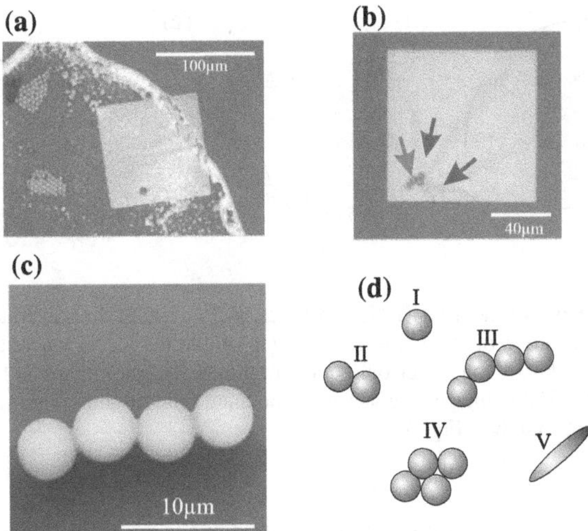

Fig. 4.2 Typical structures of micro and nano-particles mimicking cells and bacteria. **a** Deposition of 5 μm polystyrene beads using a nanoliter pipette results in severe solvent remains and uncontrolled deposition. **b** Using a printhead from an inkjet printer four 5 μm beads (*blue arrow*) and 900 nm beads (two are marked by *green arrows*) were deposited in a controlled manner. This light microscope image taken at 20x magnification already shows the limitations encountered with visible light microscopes, i.e. the 900 nm beads are barely visible. Although a conventional light microscope could in principle resolve them, a poor material contrast, as it would also be expected for biological matter, worsens the effective resolution and would need specialized techniques such as staining the cells for fluorescence microscopy. **c** SEM image of four 5 μm beads that self-arranged in a shape comparable to a *streptococci* arrangement. **d** Few examples of possible arrangements of coccus bacteria: *I* single coccus bacteria (typical diameter about 1 μm), *II diplococci*, *III streptococci*, *IV tetrad* and *V* rod-shaped bacteria

of deposited beads. Using this method, beads of different size were deposited on the same membrane (Fig. 4.2b). Full details are reported in [7].

Interestingly the polystyrene beads sometimes arrange themselves into structures (Fig. 4.2c) that compare well with those of bacteria (Fig. 4.2d) [6]. These so-called cocci arrangements can thus be mimicked by the polystyrene beads, although the rod-type bacteria cannot be easily created with them. Commercial suppliers offer plenty of rod-shaped nanoparticles, which were, however, not considered for this thesis. Identifying such structures in biological samples is of great importance for microbiology and clinical application since the arrangement of cocci can have significant influence on how the bacteria act on the human body [8]. Hence, this could be a potential application for the DIH experiments presented in this thesis.

4.1.2 DIH Results Using Coherent XUV Radiation

In the previous section the preparation method of samples consisting of polystyrene nanoparticles mimicking human cells and bacteria, and deposited on 15 nm thick silicon nitride membranes was presented. For the experiments presented in this section HHG was driven with the Ti:Sa laser described in Sect. 3.1.1. The XUV radiation was fed into the grating based setup (Sect. 3.2.1). For the experiments the 21st harmonic corresponding to 39 nm wavelength was selected by means of a pinhole with a diameter of 1 μm in the rear focal plane of the toroidal mirror. The distance between the CCD and the pinhole was set to $L_{\text{det}} + L_{\text{ph}} = 480\,\text{mm}$, which allowed the central maximum of the Airy pattern (see Sect. 3.3) to homogeneously illuminate the CCD. Typical exposure times were between $t_{\text{exp}} = 1{,}500\,\text{s}$ and $t_{\text{exp}} = 3{,}600\,\text{s}$, which resulted in holograms having a sufficient signal to noise ratio, such that they can be reconstructed using the algorithm outlined in Sect. 2.4. In order to do this, a numerically generated 1 μm pinhole illuminated with a plane wave is the starting point. The resulting field is subsequently propagated to the sample. As a first estimate of the sample a blank square aperture having 100 μm by 100 μm lateral dimensions is plugged in. The remainder of the algorithm performed exactly as depicted in Fig. 2.9, with the intermediate plane being $L_{\text{int}} = 20\,\text{mm}$ behind the sample, i.e. well into the far-field. The effective numerical aperture for all DIH experiments presented in this thesis (Eq. 2.50) is about $\text{NA}_{\text{eff}} \approx 0.023$, while only the magnification ζ was slightly changed due to optimizing the distance between sample and pinhole (see discussion below Eq. 2.51). Sufficient sampling of the fringes was, however, maintained in all experiments. Thus, the theoretical resolution for all DIH experiments is $\Delta r \approx \frac{\lambda}{2 \cdot 0.023} = 0.83\,\mu\text{m}$ according to Eq. 1.1.

4.1.2.1 DIH on a Si_3N_4 Membrane Covered with Dirt

The first sample imaged using DIH is depicted in Fig. 4.3a and features large areas of solvent residue across the whole membrane in addition to one 5 μm bead. The hologram was recorded (Fig. 4.3b) with the sample being $L_{\text{ph}} = 4.8\,\text{mm}$ distanced from the pinhole and an exposure time of $t_{\text{exp}} = 1{,}500\,\text{s}$. Hence the magnification (Eq. 2.51) is $\zeta = 100$. 4 by 4 pixels on the CCD were binned into one superpixel. Due to strong absorption across large parts of the membrane the measured hologram exhibits a poor signal to noise ratio. However, the algorithm outlined in Sect. 2.4 together with the modification for highly absorbing samples was able to retrieve the phase leading to the simulated hologram[3] depicted in Fig. 4.3c. The features compare well, however, the overall impression appears a little washed out, which is probably due to the noisy data. From that phase retrieval a reconstructed image of the object is obtained (Fig. 4.3d) which compares well to the microscope image. The 5 μm bead is well resolved. Apart from that, the parts of the membrane are discernible, where the solvent remains were so thick that the membrane became opaque at the corresponding position. That is mainly the case in the dirt ribbon on the right upper corner of the sample. The large area around the center of the membrane is covered by a thin

[3] With *simulated hologram* $|U_{\text{det}}(x, y)|^2$ at the end of the reconstruction is denoted. Visual and numerical comparison with the measured hologram $I(x, y)$ allows to judge whether or not the reconstruction was successful.

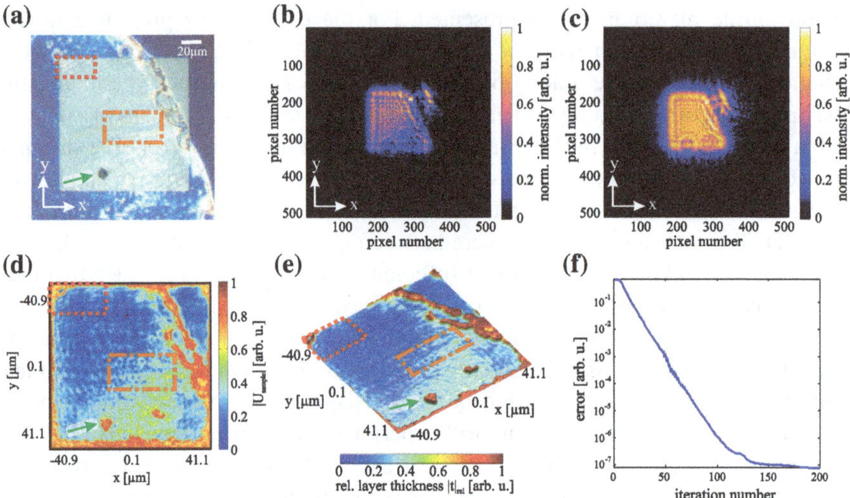

Fig. 4.3 DIH reconstruction of a contaminated silicon nitride membrane. **a** Light microscope image of the sample. The membrane is covered with absorbing solvent remains, only one 5 μm bead (*green arrow*) was deposited. Hence, the measured hologram (**b**) has a poor signal to noise ratio, note the linear intensity scale. **c** Depicts the recovered hologram after 200 iterations. The main features of the hologram were recovered. By back propagating (see Sect. 2.4 for details) the recovered hologram and subtracting the incoming illumination field one recovers an image of the sample (**d**). Comparing to (**a**) one finds the parts where the membrane is totally opaque for the XUV light, i.e. the dirt ribbon on the right upper corner and the 5 μm bead (*green arrow*). The parts in between are partially transparent indicated by the color coding (*blue* and *red* correspond to low and high absorption respectively). **e** Assuming a constant absorption coefficient allows to deduce the layer thickness and to plot it in 3D, see text. Please note that for a relative layer thickness indicated by $t_{\mathrm{rel}} \approx 1$ the layer is opaque for the used wavelength, hence this is a kind of upper cut-off for this analysis. **f** The error of the reconstruction falls very quickly and the reconstruction stagnates after about 100 iterations. The *dotted red* and *orange dash-dot boxes* mark areas which are compared in the text

layer of residue. There partial transmission occurs, which is indicated in Fig. 4.3d by the colors. A high transmission is indicated by blue colors, i.e. the incoming wave equals the outgoing wave according to Eq. 2.46. From this one can see the benefits of the XUV DIH method compared to light microscopy, namely obtaining additional information about the sample, such as layer thicknesses or homogeneity of a layer. A further benefit is the high material contrast. If a linear absorption coefficient α is assumed, one can deduce the layer thickness t by using the Beer-Lambert law and the fields introduced in Sect. 2.4 by

$$t \propto -\frac{1}{\alpha} \log \left(\frac{|U_{\mathrm{int}}(x, y) \otimes h(x, y; -L_{\mathrm{int}})|^2}{|U_{\mathrm{ph}}(x, y)|^2} \right). \tag{4.1}$$

Since α at the used wavelength is unknown, only $|t|_{\mathrm{rel}} \propto t$ can be plotted as a 3D plot, see Fig. 4.3e. Moreover, α for the bead and the dirt is certainly different. Obviously the thickness cannot be deduced at positions where the sample is opaque, hence, the height profile can only be extracted to a certain level. However, to illustrate the advantage of DIH it is sufficient at this point of the work. Analyzing some details one can see a swelling up of the thin silicon nitride due to surface tension induced by the remaining dirt film (the red dotted and orange dot-dash boxes in Fig. 4.3a, d, e) which is reproduced in the holographic reconstruction. These are surface changes which might be in the range of a few 10 nm. Unfortunately these ultra-thin membranes are too fragile to verify this with an atomic force microscope for example. The reconstruction itself converges quickly as depicted by the error function (Fig. 4.3f), which is basically computed by subtracting the measured and reconstructed hologram at a certain iteration. Please note the logarithmic plot in Fig. 4.3f. Already after a few iterations only minor changes take place and after 100 iterations the numerical noise level is reached.

4.1.2.2 DIH Phase Sensing at a Broken Si_3N_4 Membrane

In Sect. 4.1.1 it was mentioned that these ultra-thin silicon nitride membranes are rugged and can even withstand water droplets from a pipette. However, sometimes one destroys them during the preparation. An example of such a destroyed membrane with most of the membrane rolled up towards the frame is depicted in the SEM image in Fig. 4.4a. The hologram (Fig. 4.4b) was measured using an exposure time of $t_{\text{exp}} = 1{,}500\,\text{s}$ and binning two by two pixels into one. Compared to Fig. 4.3b a higher photon flux proceeds through the sample. Hence, a much clearer hologram having a better signal to noise ratio, despite twofold less binning, is measured. The reduced distance between the sample and pinhole of $L_{\text{ph}} = 4.3\,\text{mm}$ was slightly shorter compared to the previous section, thus the hologram has a slightly larger lateral dimension ($\zeta = 112$). Due to the good contrast, a much quicker convergence, basically within the first few iterations (Fig. 4.4e), is observed which results in a retrieved hologram (Fig. 4.4c) that is very similar to the measured one. The retrieved object plane, i.e. the transmission function T, is now complex-valued and depicted in Fig. 4.4d. The hue encodes the phase according to the inset color wheel and the brightness encodes the amplitude. In order to have a better contrast in the image, T was set to zero where $|T| \geq 0.3$. That corresponds to the central part of the membrane, which compares to an open aperture. The remaining silicon nitride consists of flaps that wrap around the frame, which is highlighted in Fig. 4.4d by the green square to guide the eye. Since the silicon nitride membrane has a thickness of 15 nm and the experiment is run at 39 nm wavelength, one can roughly estimate that one transition through the membrane would cause a relative phase shift[4] of $\Delta\phi \approx 0.2\pi$ and two transitions would cause $\Delta\phi \approx 0.4\pi$. From the SEM image one can at least surmise how the flap has rolled up at the edge of the frame and a few regions can be identified where a single Si_3N_4 layer remains (indicated by the magenta arrows in Fig. 4.4a, d)

[4] The refractive index of Si_3N_4 is $n = 1.24 - 0.17i$ at 39 nm wavelength [5]. Hence, $\Delta\phi = 2\pi \cdot 0.24d/\lambda$ is the phase shift of a wave passing through a Si_3N_4 layer of thickness d relative to the wave propagating through the open space of the burst window in the presented case.

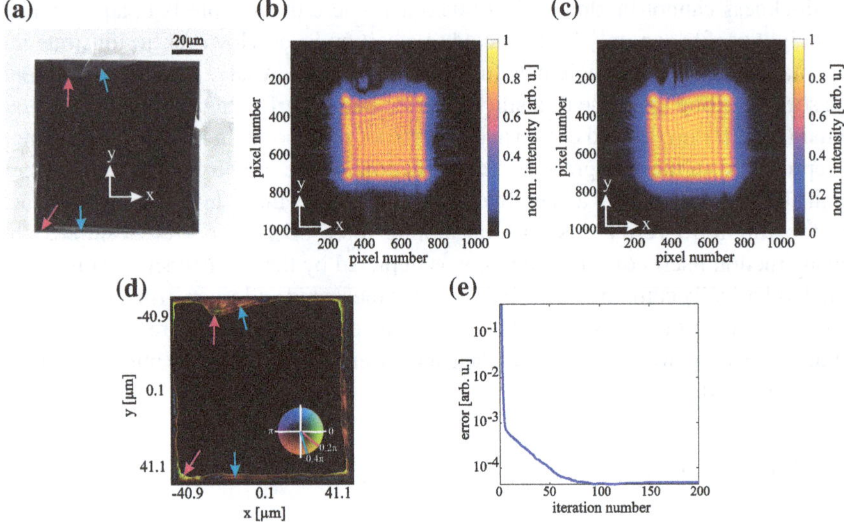

Fig. 4.4 DIH reconstruction of a destroyed silicon nitride window. **a** SEM image of the burst silicon nitride membrane. Only flaps of the membrane are left along the frame, bent over one another at some places. **b** The recorded hologram, see text for experimental details. The algorithm converges quickly and the reconstructed hologram (**c**) reproduces the features sufficiently well. In (**d**) the complex-valued transmission function of the object plane is depicted. The hue and brightness encode the phase and amplitude respectively (see *inset color wheel*). In (**a**) places featuring a single Si_3N_4 layer (*magenta arrows*) and places that indicate a double layer of Si_3N_4 (*blue arrows*), due to the flaps rolling over one another, can be found. By looking at the corresponding places in (**d**) at least an indication of the corresponding phase jumps can be found, see text. **e** The error drops already after a very few iterations indicating a good quality of the measured data

and where the flap is folded over itself, i.e. two silicon nitride layers (indicated by the blue arrows). Inspecting the phase in Fig. 4.4d one can at least roughly detect this by looking at the yellow-greenish (phase shift $\Delta\phi \approx 0.2\pi$) and dark yellow coloring (phase shift $\Delta\phi \approx 0.4\pi$) at the corresponding positions. This indicates the phase relative to the undisturbed field propagating through the center of the window.

The resolution obtained in this experiment prevents complete resolving of those flaps, but again it is sufficient to demonstrate the principle and prove that features hidden for visible light microscopes can be acquired. One of the main problems is that, as mentioned before, one layer of 15 nm silicon nitride has a transmission of a little less than 50 %, hence the flaps folded over exhibit not just an additional phase shift but also a much weaker amplitude of the transmitted light field. Hence, the limited contrast in the reconstruction.

4.1.2.3 DIH of Cocci-Like Samples on a Silicon Nitride Membrane

The main task envisioned in this section about digital in-line holography was, however, the imaging of cocci-like structures. In Sect. 4.1.1 it was explained how good samples of polystyrene beads corresponding to cocci in shape and size were prepared. In Fig. 4.5a the sample actually used is depicted. It features two 5 μm beads arranged like *diplococci*. Only small amounts of solvent remains are scattered around them and can barely be seen in the light microscope (marked by the orange arrow in Fig. 4.5a). Furthermore, a 900 nm bead is placed close to the two larger beads (marked by a green arrow). A tiny dust particle is situated almost centered on the membrane. The pinhole and the sample were brought together as close as possible. Due to the mounts of both, this distance was limited to $L_{\mathrm{ph}} = 1.9\,\mathrm{mm}$, which was somewhat perfect for this experiment because the 100 μm by 100 μm frame of the silicon nitride window was illuminated almost exactly with the same part of the Airy pattern as the CCD. The magnification is $\zeta = 253$. The hologram covers the full CCD due to this projection, or in other words, the field of view is adapted to the full size of the silicon nitride membrane.[5] The measured hologram is depicted in Fig. 4.5b; it was recorded within $t_{\mathrm{exp}} = 3{,}600\,\mathrm{s}$ using the CCD's full resolution. The reconstruction converges quickly, typically within two or three iterations, and reproduces the hologram in good detail (Fig. 4.5c). Inspecting the reconstructed object plane (Fig. 4.5a) one finds the two beads sufficiently resolved. Furthermore, the dust particle in the center is evident. If zoomed-in further (magenta square in Fig. 4.5d), one even finds the 900 nm bead reconstructed (green arrow) and also the spurious dirt can be seen (orange arrow). It is worth noting that the 900 nm bead is slightly larger than 900 nm if properly lined out in the calibrated reconstructed image, which gives a hint that the resolution limit of the system was reached approximately. This compares well to the resolution of the system that was estimated to be $\Delta r \approx 0.83\,\mu\mathrm{m}$ by the NA. In the bottom left part of the reconstructed sample (Fig. 4.5d) an increased absorption (greenish-yellowish area) is reconstructed, which seems not to be real, since no solvent was covering that part of the membrane. The assumed reason for this artifact is the illumination wave being computed by propagating the wave originating at a perfect pinhole to the sample. In the measured hologram (Fig. 4.5b) it can be seen that the membrane was not homogeneously illuminated (darker region on the bottom left part), hence the algorithm reconstructs an absorbing region there. To enhance this the actual reference wave should be measured. For future experiments this should be taken care of in order to enhance the quality of the reconstruction.

4.1.3 Summary of the DIH XUV Experiments

In the DIH section the findings about fabricating samples that feature structures having the size and shape of real-world biological objects were presented. It was found that using an inkjet printhead allows to deposit the samples onto ultra-thin silicon nitride membranes, which can then be used as targets for either CDI experiments or DIH experiments. The great benefit of XUV DIH becomes clear by looking at the

---

[5] Compare to Fig. 3.8 where the distances between pinhole and sample were larger.

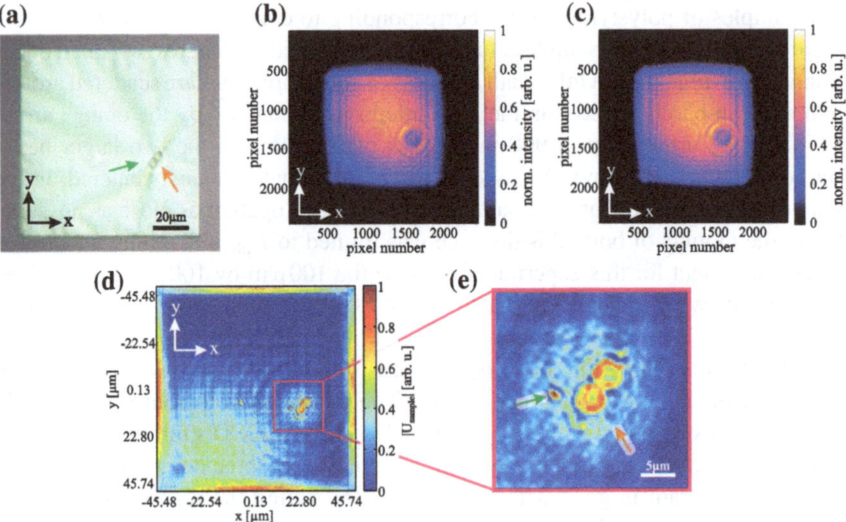

Fig. 4.5 DIH reconstruction of a *diplococci*-like sample. **a** Light microscopic image of the silicon nitride membrane with two 5 μm beads arranged like *diplococci* with little solvent remains around (*orange arrow*) and a 900 nm bead (*green arrow*). **b** The measured hologram with well resolved fringes taking full use of the CCD, see text for experimental details. The reconstructed hologram (**c**) reproduces the details sufficiently, leading to a reconstruction of the sample (**d**) at great detail. **e** A zoom-in (magenta region in (**d**)) features again the *diplococci*-like structure and the 900 nm bead (indicated by the *green arrow*). Even the small solvent scatters around the beads are resolved (indicated by the *orange arrow*). One pixel in (**d**) and (**e**) has a size of $p_{\text{obj}} = 266.7$ nm in both directions

experiments that were conducted in this work. The object support constraint, which is essential for CDI, is not needed in DIH. Moreover, the holograms can be measured with a single exposure at a low dynamic range. Note that all holograms that were presented in this section featured a maximum count rate at the brightest pixels of not more than 2,000 counts while the noise-level is at about 300 counts. Hence the dynamic range in these measurements is just about one order of magnitude, which is much less than what is needed for CDI (see Sect. 3.2.3). The ability to change the field of view continuously by changing the distance between the pinhole and the sample (Fig. 3.8) is also beneficial and allows to find a sample quickly. Moreover, it was demonstrated that the phase information can be used to assess even 3D information of the sample despite just capturing the hologram from one view. That property is inherent to holography in general. In order to assess this information one can either use the phase, which allows very sensitive depth measurements with respect to the wavelength, or one can use the typical strong absorption of solid matter for XUV light. It also worth noting that DIH experiences no depth of focus limitation compared to a conventional light microscope. As usual in holography, one can take advantage of the recovered complex-valued wave-field and propagate it to other z-positions and one can therefore get diffraction limited images for different (focal) planes. The reconstruction of the holograms is straightforward and the algorithm converges very quickly, typically after two or three iterations. The quality of the reconstructions could be further improved if one first determines the incoming illumination wave by measurement. In holography this is typically referred to as measuring the reference wave. In the presented work only a computed illumination wave was used, which, however, already allowed for remarkable reconstructions. Probing the real illumination wave and using this to reconstruct the holograms should yield an even better quality of the reconstruction.

The reconstructions done in CDI using an HIO algorithm can easily invoke well above 10,000 iterations, especially if the guided HIO algorithm is used, where many separate runs are used to enhance the result. What makes all these algorithms slow is the numerical Fourier transform, i.e. the FFT. The HIO just needs two FFTs per iteration, while the DIH reconstruction code needs, besides two FFTs, additional steps e.g. for data rescaling and the multiple multiplication of the Fresnel propagator. Moreover, the DIH reconstruction algorithm needs a much finer computation grid to maintain the phase[6] during the wave propagation. The HIO algorithm for CDI, that switches between object plane and Fourier transform plane, in comparison just needs a computation grid as large as the oversampled diffraction pattern captured in an experiment.[7] However, in this section it was demonstrated that only very few iterations are needed for the reconstruction of a digital in-line hologram, hence the necessary overall computation time is very short compared to 10,000 iterations in an HIO algorithm. Using a typical desktop computer[8] and an implementation of the DIH reconstruction algorithm in MATLAB the processing rate was at roughly one iteration per second at full resolution. This means one could easily implement the whole reconstruction process in real-time, taking advantage of fast data processing on a modern multi-core graphic card processor and a faster programming language, e.g. C++. Limiting for real-time table-top XUV DIH at the moment is the XUV photon flux, which currently keeps the needed exposure times long. However, one should note that the system used for the DIH experiments in this work was far from perfect with the 1 μm pinhole placed in a 50 μm (FWHM) focus wasting almost all of the flux. Using the FCPA system (Sect. 3.1.2) and the dielectric mirror based setup (Sect. 3.2.2) having a focal spot size matching well to the pinhole diameter, one could certainly get to exposure times in the range of a minute or even less, setting the stage for real-time high resolution XUV DIH. This should be a subject for experiments in the near future.

---

[6] For the DIH reconstructions presented in this thesis the computation grid was typically between 4000×4000 and 7000×7000 pixels large, demanding a high amount of memory.

[7] If the measured diffraction pattern in CDI is highly oversampled one can even reduce the computation grid further by binning pixels together without losing resolution in the reconstructed object, as was done in Sect. 4.3.

[8] A processor operating at 3.1 GHz having four cores and 8 GB random access memory.

4.2 Transmission CDI at the Abbe Limit with a High Numerical Aperture

Each experiment that is presented in the previous and the following sections of this chapter demonstrates the beauties and advantages of using coherent XUV light for microscopy. However, the resolution is in all cases limited to the micron range, which is definitely in reach of visible light microscopes. The best resolution that was reported for a CDI experiment so far was in the range of 1.6 times λ [9], which equals a spatial resolution of 22 nm and is thus already not far from the Abbe limit.[9] Using the fiber CPA setup (Sect. 3.1.2) producing very narrow-band harmonics together with the dielectric mirror based imaging setup (Sect. 3.2.2), recently experiments at a high numerical aperture were conducted with the objective to reach sub-wavelength resolution and to approach the Abbe limit (Eq. 1.1). The wavelength for all high NA experiments reported in this section is $\lambda = 33.2$ nm, which corresponds to the 31st harmonic of the ytterbium doped fiber laser.

The samples used were fabricated out of 200 nm thick silicon nitride membranes on a silicon substrate geometrically equal to those depicted in Fig. 4.1. They were covered with an additional gold layer of 200 nm thickness to render the membrane totally opaque for XUV light.[10] Using focused ion beam milling (FIB, FEI HELIOS NANOLAB 600I), apertures were cut into the membranes, which are hard edge planar apertures for XUV light. The results achieved will be presented in the subsequent sections.

4.2.1 High NA Reconstruction of a Pinhole Aperture

The first object measured in this part of the work was an aperture having a diameter of 1 μm (Fig. 4.6a). This rather simple object was chosen for calibrating the sample-to-detector distance, in order to accurately determine the numerical aperture. For this experiment a beam stop consisting of an approximately 100 μm large sugar crystal affixed to a 5 μm tungsten wire was used. The exposure time was $t_{exp} = 600$ s and 4 by 4 pixels of the CCD were combined into one, which results in an approximate linear oversampling ratio[11] of $O \approx 6$ (Eq. 2.25). After applying the curvature correction and intensity normalization (Sect. 3.4) further pixel binning resulted in a computation grid of 308 by 308 pixels, such that $O \approx 3.7$. The prepared diffraction pattern is depicted in Fig. 4.6b. From a transmission SEM measurement (Fig. 4.6a) the diameter of the aperture was accurately known, hence the algorithm outlined in Sect. 3.3 can be employed to determine the distance between sample and CCD more accurately. It was

---

[9] For very special sample geometries providing sparsity sub-wavelength CDI was demonstrated for visible light [10].

[10] The bare 200 nm silicon nitride has a residual transmission of 0.02 % [5].

[11] From a first estimate by mechanically measuring the distance it was clear that the distance between sample and CCD was in the order of 10 mm.

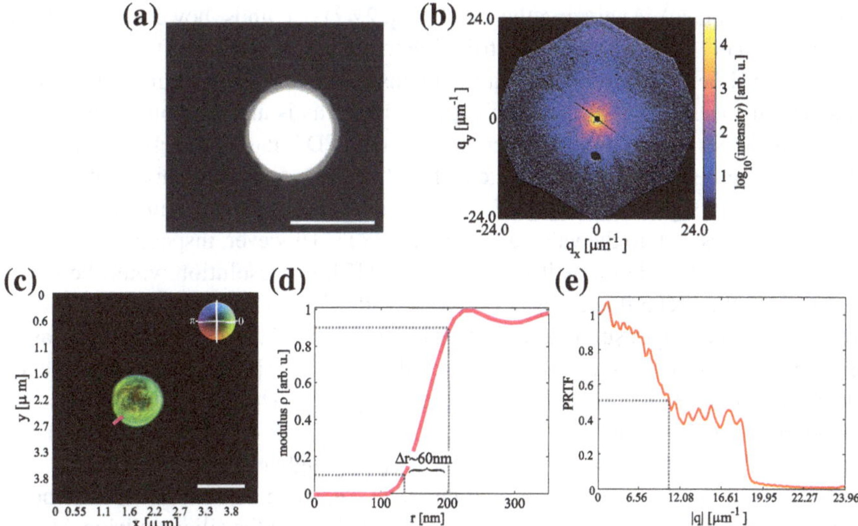

Fig. 4.6 High NA experiment using a 1 µm aperture as object. **a** A transmission SEM (STEM) image of the aperture. The FIB fabrication produced soft edges which are visible as *gray shading* in the image, see text for details. **b** The measured diffraction pattern with the curvature corrected and the intensity normalized. The NA is ≈0.80 with the sample being 10.46 mm away from the CCD. Note the beam stop blocking the central speckle of the diffraction pattern. **c** Reconstructed object plane featuring the aperture. Positive complex values were allowed, indicated by the hue and brightness, which encoded the phase and amplitude respectively. A profile (**d**) marked with *magenta* in (**c**) suggests a resolution of $\Delta r \approx 60$ nm. **e** The PRTF reveals what is also visible in (**b**), i.e. no full advantage of the high NA can be taken, since intensities were measured only to an effective NA of ≈0.36. Hence, the resulting resolution is in the order of $\Delta r \approx 70$ nm. The scale bars in (**a**) and (**c**) are 1 µm

determined to be $L_{\text{Sample-CCD}} = 10.46 \pm 0.05$ mm in this experiment. The numerical aperture is thus NA ≈ 0.80, allowing for a maximum detectable momentum transfer $q_{\max} \approx 24.02\,\mu\text{m}^{-1}$ at the midpoint of the detector's edges, resulting in a theoretical resolution limit of $\Delta r \approx 20.8$ nm, i.e. well below the wavelength.[12] The prepared diffraction pattern was fed into a guided HIO algorithm (Sect. 2.3) which performed 1,000 iterations on a set of 10 independent reconstructions that were each starting with random phases. After these 1,000 iterations the best reconstruction, judged by the error, was used to generate a new set of 10 new initial guesses as the next generation. After 20 generations a common solution was found. The shrink-wrap method was used to update the support during the iterations accordingly. In Fig. 4.6c the reconstructed object plane, which was allowed to have positive complex values, is depicted. The aperture is properly reconstructed with an almost flat phase across the aperture (the hue encodes the phase and the brightness the amplitude of the complex-

[12] *Maximum detectable momentum transfer* refers to the highest measurable modulus of the projected momentum transfer vector **q** for all angles ϕ (Fig. 3.9) with respect to the center of the CCD, e.g. at the CCD's midpoints of the edges.

valued object plane). Analyzing the PRTF (Eq. 2.42) one finds, however, that above a $|q| \approx 6.9\,\mu\text{m}^{-1}$ no significant contribution to the reconstructed diffraction pattern occurs (Fig. 4.6e).[13] That means that the intensities measured at higher momentum transfers, if measured at all, are subject to noise. This is also obvious in Fig. 4.6b, where just noise is visible in the outer parts of the CCD image. The resolution is thus effectively limited to $\Delta r \approx 70\,\text{nm}$ due to this effective NA. Testing this resolution by taking a profile (Fig. 4.6d)[14] in the reconstructed object plane results in $\Delta r \approx 60\,\text{nm}$, which compares well to the value given by the PRTF. However, inspecting the PRTF and using a $1/e$ threshold, as it is done e.g. in [12], the resolution would be much better, since the PRTF fluctuates between 0.4 and 0.5 until $|q| \approx 14.5\,\mu\text{m}^{-1}$, which would in turn already result in a resolution equivalent to one wavelength. The reason as to why the resolution is not much better could be that the aperture has a soft edge for the XUV, as can be seen by the gray shading in Fig. 4.6a. The aperture was milled by shooting gallium ions onto the gold covered silicon nitride from the wrong side, i.e. from the gold side. Because the Ga ions have a much higher interaction cross-section with gold one gets a kind of under-etching effect, i.e. some gold around the intended aperture was removed before the ions could break through the silicon nitride. Hence in Fig. 4.6a black colors correspond to Au on Si_3N_4, gray corresponds to a beveled Si_3N_4 membrane, and white to a clear aperture. This explains why the edges in the reconstructed object plane appear washed out: the remaining beveled Si_3N_4 acts as a kind of soft edge. For the next presented sample the aperture was milled from the other side resulting in a sharp edged aperture, see Fig. 4.7a.

4.2.2 High NA Reconstruction of a Complex-Shaped Aperture

Since the circular aperture used in the previous subsection is a somewhat trivial sample, another more complex sample had to be imaged and will be dealt with in this section. The sample used is depicted in a transmission SEM image in Fig. 4.7a. Please note that this time the FIB milled the sample from the silicon nitride side, which produced a hard gold edge at the exit plane facing the CCD. The distance of the sample to the CCD was kept constant at 10.46 mm, as for the previous sample. The diffraction pattern was measured by combining single measurements with exposure times $t_{\text{exp}} = 1800, 600, 300, 60\,\text{s}$, while the two shortest exposures had the beam stop removed. Hence, a high dynamic range (Sect. 3.2.3) diffraction pattern could be fed into the same algorithm as used in the previous subsection after applying curvature correction and intensity normalization (Fig. 4.7b). Again the object was retrieved at a convincing quality (Fig. 4.7c). The phase is flat across the object[15] and one can

[13] The conservative value of PRTF = 0.5 as threshold is used.

[14] The 90 %/10 % of the modulus of the electron density ρ criterion is used to determine the resolution of the experimental data [11].

[15] The different hue is due to a constant phase offset compared to Fig. 4.6c, which is one of the remaining ambiguities in a CDI reconstruction.

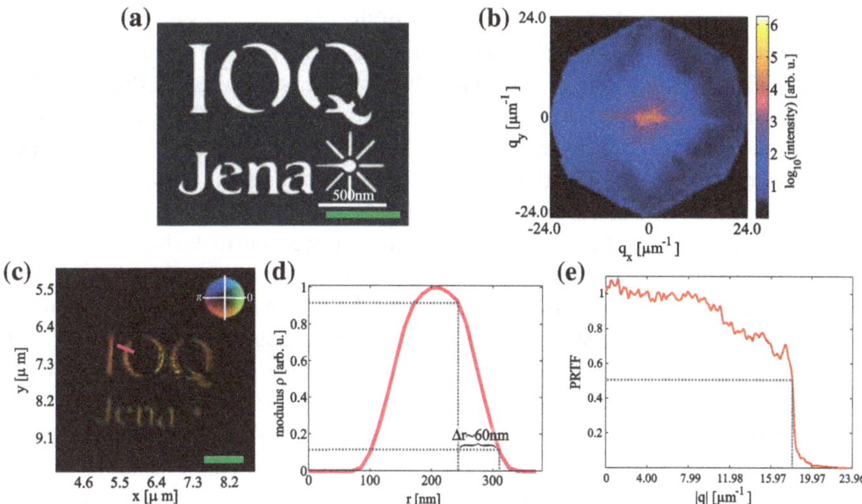

Fig. 4.7 High NA experiment using a complex-shaped aperture. **a** A transmission SEM (STEM) image of the aperture. In contrast to Fig. 4.6a a hard gold edge is evident. **b** The measured diffraction pattern composed of a high dynamic range diffraction pattern from several single exposures with and without a beamstop. The curvature was corrected and the intensity normalized. The NA is ≈0.80 with the sample being 10.46 mm away from the CCD. **c** Reconstructed object plane featuring the object. Positive complex values were allowed, indicated by the hue and brightness, which encode the phase and amplitude respectively. A profile (**d**) at the position marked with the *magenta line* in (**c**) suggests a resolution of $\Delta r \approx 60$ nm. **e** The PRTF shows that a much more usable diffraction signal compared to Fig. 4.6b was measured. The resulting resolution is well below one wavelength. The *green scale bars* in (**a**) and (**c**) are 1 μm

easily identify the parts of the object in comparison to the STEM image (Fig. 4.7a). Unfortunately, the tiny rays of the spark in the bottom right corner of the object were not resolved. Also the fine details in *Jena* are not resolved well. These are well below one wavelength if accurately measured out in Fig. 4.7a, hence an increased flux would be necessary to get some light through these tiny features. Taking a profile yields the experimentally achieved resolution, which is again in the range of $\Delta r \approx 60$ nm (Fig. 4.7d). Comparing the measured diffraction pattern with the one in Fig. 4.6b one finds that the fringes extend more towards the edge of the detector. Hence, a better resolution could be expected. By analyzing again the PRTF (Fig. 4.7e) an upper limit of diffraction data that contributed to the reconstruction is found at $|q| \approx 18.06\,\mu\text{m}^{-1}$, which compares well to a visual inspection of the measured diffraction pattern. The resulting experimental resolution is thus $\Delta r \approx 27.6$ nm, i.e. equivalent to 0.8 wavelengths.

This result is a bit puzzling, because for the round aperture in the previous subsection the experimental resolution obtained from both, a profile and the PRTF, compare well, especially considering the soft edge. Here the complex-shaped aperture has a defined hard gold edge, thus this should not limit the resolution. On the other hand, if the resolution of around 27 nm would be real, one should start to see the rays of the spark in the bottom right edge of the object. However, even when the reconstruction algorithm is seeded with a perfect fitting support deduced from the transmission SEM image (Fig. 4.7a) none of the sparks arise. Therefore, the resolution that is reached practically is in the range around $\Delta r \approx 60$ nm, i.e. 1.8 wavelengths, as suggested by the profile measurement. That value also fits better to a visual inspection of the reconstructed object plane, which also features some distortions, e.g. in the letter Q. As discussed for instance in [13, 14] another possibility would be that fringes from the inner surface of the aperture are measured due to the high NA. Thus a full 3D reconstruction would be necessary to achieve the resolution suggested by the highest measured momentum transfer. This would also explain the distortions on the larger structures within the structure, e.g. the letter Q, because these would be the first from where one could expect those scatters from the inner surface due to the larger aperture opening compared to the smaller structures. Consequently, implementing a 3D reconstruction code, e.g. the ankylography code [15], would be beneficial in the near future. Another option as to why the expected resolution is not achieved could be missing temporal coherence. However, as discussed in Sect. 3.1.2, the temporal coherence of the system used in this work should be sufficient for sub-wavelength imaging. Moreover, a low temporal coherence would blur the reconstruction and not cause these additional sharp edged features as they appear in the letter Q. Hence, the finite depth of the sample seems to effectively limit the achievable resolution.

4.2.3 Summary of the Transmission CDI Experiments

In this section very recent and to some extent preliminary results were presented. The first experiment using a 1 μm aperture was used to determine the exact distance between the sample and the CCD by supplying the defined pattern of such an aperture and the exact knowledge of the diameter. The achieved resolution in terms of the effectively reached NA compares well to an estimation based on taking a profile in the final reconstruction. It was demonstrated that the aperture can be reconstructed to a resolution of approximately 60 nm, which is equivalent to 1.8 wavelengths. As a more complex sample an aperture featuring parts of the institute's logo, where this work was done, was used. Astonishingly, the object could be recovered showing almost all features in good detail. One should keep in mind that all the reconstructions were run by means of a guided HIO algorithm without any a priori knowledge about the sample. The resolution achieved is in the same range as for the aperture sample, despite the measured diffraction data suggesting an equivalent resolution slightly below one wavelength. It was further found that the finite depth of the sample likely causes the resolution limit, such as is reported in literature for very high NAs [14]. Hence, for future experiments a full three-dimensional reconstruction should be done in order to achieve real sub-wavelength resolution. Moreover, for a better confidence in the data, an increased flux from the HHG is needed and will likely be implemented in the near future.

4.3 CDI of an Artificial Non-periodic Specimen in Reflection Geometry

The sample that was investigated in reflection geometry[16] consisted of the characters *V 20* fabricated by electron lithography on a silicon wafer (Fig. 4.8a). The height of the gold layer from which the structure was etched was 30 nm, which corresponds to a little less than one wavelength ($\lambda = 38$ nm) of the illumination used. The overall dimensions of the sample are roughly 33 μm by 11 μm with details of the structure being in the sub-micron level. The sample was placed in the rear focal plane of the grating based CDI setup (Sect. 3.2.1) and HHG was driven with a Ti:Sa laser (Sect. 3.1.1). A 50 μm pinhole was inserted closely before the focus in order to select the 21st harmonic at $\lambda = 38$ nm wavelength. Please note that in contrast to the calibrated spectrum of the source that was presented in Fig. 3.2 the wavelength used in this section was measured by the pinhole method—having a 5 μm pinhole temporarily installed—described in Sect. 3.3. The small deviation could be subject to a blue shift of the HHG spectrum due to a variable day-to-day laser performance. However, the small deviation has no effect on the results presented here, thus $\lambda = 38$ nm will be used for the evaluation in this section. The angle of incidence was aligned to roughly $\alpha_i \approx 22.5°$ and the diffraction pattern was captured 430 mm downstream from the sample using a CCD (2048 × 2048 pixels, each 13.5 by 13.5 μm wide) cooled to −70 °C in order to suppress thermal noise. This geometry corresponds to a numerical aperture NA ≈ 0.03 and results in the momentum transfer component mapping that was depicted in Fig. 3.12a–c. The linear oversampling ratio (see Eq. 2.25) in this geometry becomes $O \approx 37$ if the full CCD resolution is assessed. As discussed in Sect. 2.2.3, an O of two is sufficient for a phase retrieval. Thus, one can safely bin 4 × 4 pixels of the CCD into one super-pixel in order to reduce the exposure times without losing the necessary oversampling. The binned raw diffraction pattern that was measured ($O \approx 4.6$) is depicted in Fig. 4.8b and consists of an HDR diffraction pattern (see Sect. 3.2.3) combined from seven single measurements having $t_{\text{exp}} = 3, 6, 12, 24, 48, 96, 192$ s exposure times. Thus, the diffraction pattern is strongly enhanced in terms of the dynamic range. The center of the beam was blocked by a 2 mm large beam stop, which was placed close to the CCD. All hotspots were carefully removed and the region of the beam stop was set to zero.

Inspecting Fig. 4.8b one finds that the diffraction pattern is slightly off-center. From the properties of the Fourier transform it is known that a translation of the diffraction pattern by $\mathbf{q}_0$ results in a phase ramp $\exp(i\mathbf{q}_0\mathbf{x})$ in the object plane image. This effect was visualized in the final reconstruction in the scheme depicted in Fig. 2.7b, where the pattern was intentionally shifted off the center. If one knew this shift exactly one could account for it. However, it is of course more convenient though to simply center the measured diffraction pattern as well as possible. This is even more important if additional constraints, such as positivity (see Sect. 2.3), are to be imposed on the object plane data. Moreover, it is known from the momentum transfer

[16] Please refer to Fig. 3.11 for the coordinate definition in this section.

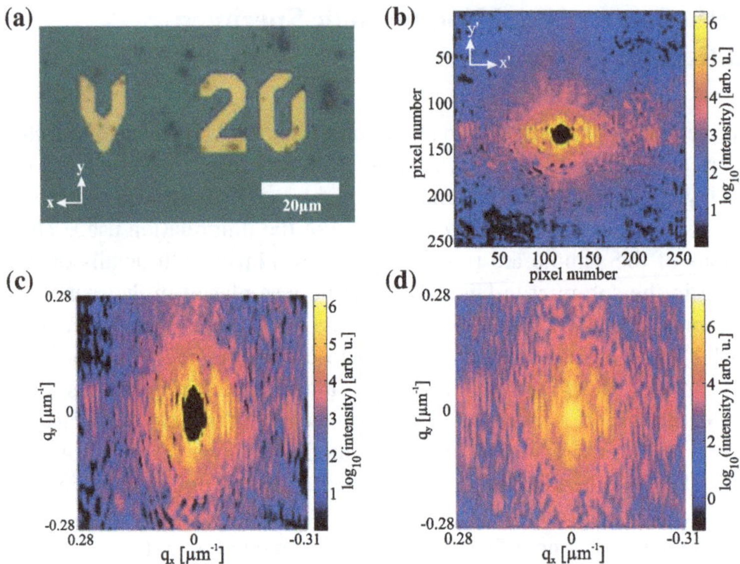

Fig. 4.8 Experimental diffraction pattern obtained from a non-periodic specimen in reflection geometry. **a** Light microscope image of the sample, see text for geometric details. Note the dirt on the sample, which introduces additional distortion. **b** Depicts the measured HDR diffraction pattern, see text for experimental parameters. 4 by 4 CCD pixels were binned to one superpixel, hence the measured grid is 256 by 256 pixels. The *bright center* of the diffraction pattern was blocked with a beam stop. **c** By centering the diffraction pattern and stretching it in the *vertical axis*, in order to have an almost identical scaling of the q_x and q_y component, the diffraction pattern (**c**) that is fed into the HIO algorithm is prepared. **d** Shows the reconstructed diffraction intensity, i.e. the square of the modulus of the Fourier transform of the reconstructed object, after 10,000 HIO iterations. The unknown features, which were blocked by the beam stop, are satisfyingly recovered. Compare to Fig. 3.11 for the coordinate systems' and axis' definitions

vector discussion for reflection geometry (Sect. 3.4.2 and Fig. 3.12a–c) that the q_x and q_y components do not scale equally on the planar quadratic detector. At this low NA the curvature of the Ewald sphere is negligible. The q_z component of the momentum transfer vector can be neglected as well, since the sample is sufficiently flat. However, the fact that the detector observes the sample under an angle of 22.5° leads to an unequal momentum transfer scaling in the x' and y' direction of the detector. Due to the low NA q_x and q_y sample almost uniformly and independently across the detector, and the phase retrieval should result in a horizontally reduced image of the sample. Unfortunately, all attempts to reconstruct the sample in that way failed. Hence, the measured diffraction pattern was rescaled such that q_x and q_y scale almost equally across the detector. This is done by stretching the diffraction pattern in the detector's y' direction by a factor of $1/\sin(22.5°) = 2.61$. This corrected diffraction pattern is depicted in Fig. 4.8c.

For the phase retrieval the HIO algorithm, as outlined in Sect. 2.3, with a feedback parameter β ramped from 3 down to 0.1 within 10,000 iterations was used. After 90 iterations of the HIO algorithm another 10 iterations using the error-reduction algorithm were performed. Further, positivity was enforced, i.e. positive complex-valued numbers in the object plane. The reconstructions started with random phases and an initial support estimated from the autocorrelation of the measured diffraction pattern [16]. During the reconstruction the shrink-wrap method [17] was applied to update the support every 10 iterations. The Gaussian blurring that is necessary to estimate the new support in shrink-wrap was also ramped down from a 10 pixels standard deviation radius to a one pixel standard deviation radius in the end. For all pixels where no data was measured, i.e. mainly the beam stop region, the amplitude and phase evolve freely, while for all measured pixel values the amplitude constraint in the detection plane was enforced.

Using this method some good reconstructions were achieved, while it is worth noting that the success rate of the reconstructions was on the order of $\approx$80 %.[17] Mostly, the reconstructions did not converge when shrink-wrap shrunk the support too quickly at a certain step, which eventually leads to a complete run-away of the phases and instability of the algorithm. An example of a successfully reconstructed diffraction pattern, i.e. the square of the modulus of the Fourier transform of the reconstructed object, is depicted in Fig. 4.8d. One sees easily that all measured features are contained and that the unknown region behind the beam stop was successfully retrieved from the remainder of the pattern without any additional knowledge. One missing feature in Fig. 4.8d, compared to Fig. 4.8c, is the missing ring structure[18] that is evident in Fig. 4.8c. This is caused by a slight diffraction at the 50 µm pinhole that is used to select the single harmonic. However, the diffraction pattern caused by the object is dominant. The scaled object plane is depicted in Fig. 4.9a and compares well to the actual object. The reconstructed phase across the object is relatively flat, as was to be expected from the planar letters. The main features of the object are resolved to satisfaction. Actually the object is reconstructed mirrored in the computer due to the reflection geometry, however, for convenience the result presented here was numerically mirrored in the final step to allow a comparison with the light microscope image in Fig. 4.8a. Moreover, one sees that the reconstructed object plane now scales identically in the x and y direction, which is due to the stretching of the diffraction pattern.

The phase retrieval transfer function (PRTF, Eq. 2.42) can be used to determine the achieved resolution. For this experiment the PRTF is depicted in Fig. 4.9b. Apart from the undefined part that is blocked by the beam stop (red bar in Fig. 4.9b), the PRTF stays well above the 0.5 level (dotted line). If the PRTF drops below 0.5 [1], or less conservative $1/e$ [18], the corresponding momentum transfer would suggest the resolution obtained in a particular experiment. Thus, in the experiment presented the resolution is only limited by the low NA. One can compute the resolution limit[19] to $\Delta r = \frac{1}{2q_{\max}} \approx 1.7\,\mu\text{m}$. This compares well to a qualitative inspection of the

[17] *Success* in this case is measured by the error metric, which was mentioned in Sect. 2.3.

[18] More precisely these are ellipses in the stretched pattern.

[19] Note that the NA should in principle allow a resolution of $\Delta r \approx 0.63\,\mu\text{m}$. This is effectively reduced here due to the projection.

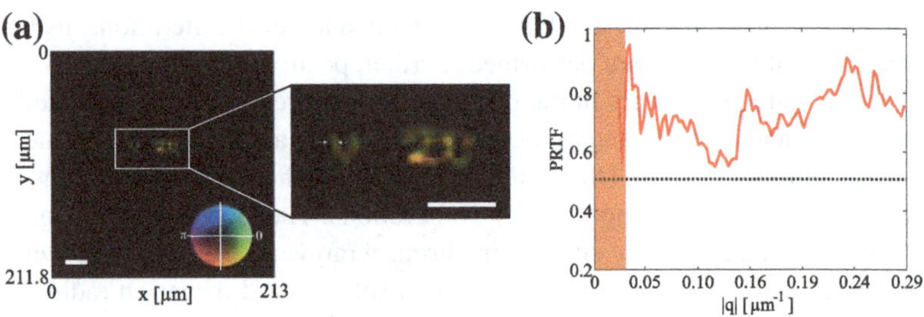

Fig. 4.9 Reconstruction of a micron scale object in reflection geometry. **a** The reconstructed complex-valued object plane. The brightness encodes the amplitude of the object and the hue encodes the phase (see *inset colorwheel*). Inspecting the phase, one finds that it is relatively flat across the object. The resolution can be estimated from looking at the *vertical bar* of the letter *V* (*white arrows*) in the magnification which is 3 μm wide and well resolved. The scale bar is 20 μm. A more quantitative access to the resolution can be gained from the PRTF depicted in (**b**). It shows, by not dropping below 0.5 for all values $|\mathbf{q}|$, that the resolution is only limited by the NA and maybe noise. Thus the calculated resolution is roughly 2 μm which compares well to the qualitative inspection in (**a**)

reconstructed sample (Fig. 4.9a), where a well resolved vertical bar of letter *V* (marked by two white arrows) corresponds to a length of 3 μm. Thus, the achieved spatial resolution is about 2 μm.

The results obtained in reflection geometry presented in this section were recently published [19]. In [19] additionally the effect of imperfect centering of the diffraction pattern by shifting it pixelwise around and running the reconstruction is discussed. The result is that shifting the diffraction pattern as little as 1 pixel off the perfect center spoils the reconstruction severely. This is mainly caused by the phase ramp that is introduced, which causes 2π phase wraps almost from one pixel to the next which are then tackled by the positivity constraint. Moreover, different vertical stretching factors were tested. The result was that the reconstruction produces faulty results, if converging at all, when the stretching deviates significantly from the perfect value, i.e. stretching by a factor of 2.61 induced by the angle of incidence. This is not intuitive, because a stretched object results in a reduced diffraction pattern on the corresponding axis in Fourier transform plane, provided that the sampling distance is kept constant. Hence, in principle, one should be able to reconstruct the stretched object plane without altering the data. The behavior found is explained as follows. The whole reconstruction process takes place in the object plane and the detection plane, i.e. the Fourier transform plane. The only definite information is known in the Fourier transform plane, i.e. the measured diffraction pattern, while almost nothing about the object plane is known, except the rough bounds due to the autocorrelation of the diffraction pattern. Martin and co-workers convincingly discussed the effect of noisy data on the HIO algorithm [20] and found that already small amounts of noise hinder convergence. If one has a stretched object it will cover more pixels in the object plane and, hence, the area of the support is bigger compared to the unstretched object. This also means that more noisy pixels are carried on from iteration to iteration, because the noise in the recorded diffraction pattern, which comes from experimental measurements and not from the object, is projected onto the object plane at every step [20]. Hence, minimizing the area covered by the object while keeping the computation grid fixed reduces the effects of noisy data. This could be the main reason as to why the reconstructions for the presented experiment succeeded once the data was stretched in the way that a 1:1 scaled image appeared in object plane. Interestingly, this is also beneficial the other way around, such as Sun et al. presented in [21], where they used a strongly stretched object in reflection geometry, and found that the stretching factor of the object (their institute's logo) corresponds to the stretching observed by the detector due to the tilted sample plane. In their reconstruction using HIO they scaled back the logo to 1:1. Compared to [14] the experiment presented has achieved a much lower resolution in reflection geometry. However, the quality of the reconstruction, bearing in mind the complexity of the object, is much better. This clearly indicates that an increase of the NA also increases the trouble with properly preparing the experimental data.

The resolution obtained in the experiment presented here is relatively low ($\Delta r \approx 2\,\mu\text{m}$) compared to state-of-the-art light microscopes. Nevertheless, it is rather high if one considers the large distance between the sample and the detector, which is almost half a meter. Compared to a conventional light microscope one would need to get as close as a few millimeters to the object in order to achieve an equivalent resolution. Here the clear distance is 430 mm from the sample to the detector, hence, one could think of implementing additional diagnostics such as a spectrometer to measure fluorescence, which would open the way to multimodal microscopy. Furthermore, in the CDI approach the phase of the object is measured as well. Hence, one can measure very precisely height profiles as long as the heights remain smaller than the wavelengths used, because otherwise an unavoidable phase wrap would induce ambiguity. Also, by choosing different illumination wavelengths out of the harmonics spectrum, one could assess material [22] and phase contrasts [23].

4.4 CDI Based Fast Classification of Breast Cancer Cells

After having investigated artificial samples in the preceding sections, the focus of this work will now turn onto the application of the findings for clinical purposes. For instance in cancer treatment it is highly desirable to identify the characteristics of cancer as fast as possible, since quick proper treatment is essential for a successful therapy and for reducing the risk of spreading malignant cells via the blood or lymph system. Nowadays, the standard method to identify the cancer characteristics is profiling by polymerase chain reaction (PCR) or sequencing. In daily clinical routine, patient samples are usually sent out to external labs for an analysis and thus round-trip times from the tissue extraction until report reception are several days. Hence, the need for faster reliable on-site classification methods with minimal need for sample preparation is evident. A good overview about the clinical application of PCR analysis can for instance be found in [24].

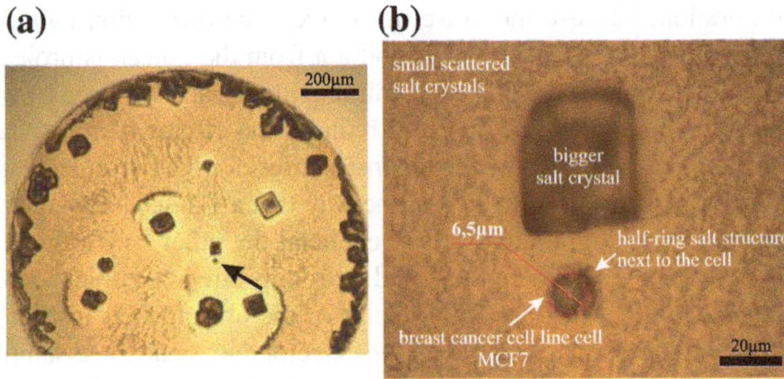

Fig. 4.10 Microscopic images of a dried PBS droplet containing a cancer cell. **a** An overview over a dried out droplet applied with a micropipette. The cell is marked with the *arrow*, all other objects of the droplet are crystalline remains of the PBS buffer. **b** At high magnification one MCF7 cell with a diameter of roughly 6.5 μm and plenty of crystalline salt remains are found. The cell has an adjacent half-ring of salt

The analysis of Raman spectra of cells and bacteria has been under intense investigation during recent years [25, 26] to approach this problem. In this thesis a different approach to classify rapidly cell types based on coherent diffraction imaging techniques is presented. The presented method [27] could allow an implementation in a clinical environment. In a proof-of-principle experiment, differentiation between different breast cancer cell lines of the same cancer type but with different expression profile and different genetic aberrations was attempted. The cell lines used are named MCF7 and SKBR3. These are epithelial-like cell lines, which were established from pleural effusions of a 69-year-old and a 43-year-old Caucasian woman with metastatic mammary carcinoma, respectively. The cell line cells are commercially available and are well-established as a good model to mimic circulating tumor cells in the blood stream of breast cancer patients. These unstained and unlabeled cells were cultivated and afterwards fixed in a liquid PBS buffer. Using an inverted microscope and a micropipette, droplets containing several cells were deposited onto plane substrates, which were previously covered with a 200 nm gold layer to enhance XUV reflectivity.[20] After drying one typically finds large areas scattered with crystalline salt remains from the PBS buffer and some cancer cells in between. An overview and a detailed microscopic image of the deposition result are depicted in Fig. 4.10.

For the experiments the Ti:Sa based laser source was used to generate coherent XUV radiation (Sect. 3.1.1), which was subsequently fed into the grating based CDI setup (Sect. 3.2.1). A 5 μm pinhole was employed in the rear focal plane of the toroidal mirror in order to select the focus of the 21st harmonic at 39 nm wavelength. The distance from the pinhole to the CCD was $L_{\text{obj-CCD}} \approx 365$ mm in reflection geometry (Fig. 3.11) using an angle of incidence $\alpha_i \approx 22.5°$. Hence, one can think

[20] Carbon and gold have a comparable reflectivity at used wavelength and angle of incidence [5]. Hence, one probes the surface and silhouette of the specimen at the same time to acquire a maximum of spatial information.

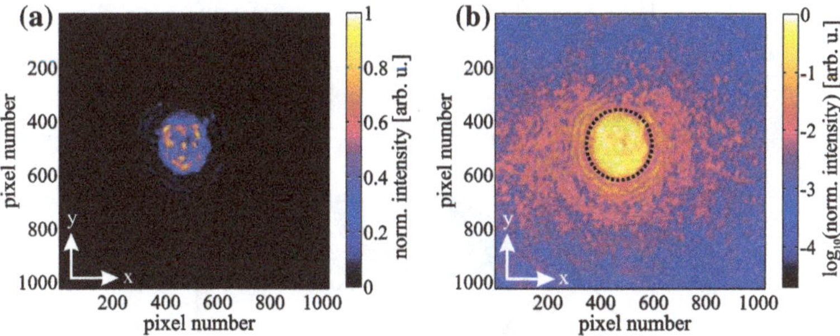

Fig. 4.11 Example of a raw diffraction pattern captured in reflection geometry on a MCF7 cell. **a** Linear intensity plot of the raw diffraction data mainly featuring the hologram part in the *center*. **b** A logarithmic intensity plot of the same data shows plenty of CDI related fringes all over the detector. The *dotted circle* approximately delimits the hologram from the CDI related fringes

of the geometry as a combination of the reflection CDI experiment described in Sect. 4.3 and the DIH experiments in Sect. 4.1.2. The sample was placed as close as 0.5 mm to the pinhole, which means essentially that the illumination from the pinhole allows to access the diffraction pattern of a single cell. Following the discussion in Sects. 2.4 and 4.1.2 one finds a hologram in the central part of the detector, featuring an extremely poor spatial resolution compared to the DIH measurements in Sect. 4.1.2. The remainder of the detector will be covered with coherent CDI related speckle caused by the silhouette and surface of the cell.[21] This geometric arrangement is a mixture of CDI and DIH and compares to what is known in literature as *Fresnel CDI* [28]. However, here it is combined with reflection geometry and a compact table-top HHG source. The advantage over pure CDI, i.e. using a much larger pinhole and hence illuminate the sample with a pure plane wave, is that the strong central speckle of the diffraction pattern is suppressed and thus neither a beam stop nor multiple exposures are needed to collect data with a high dynamic range. In Fig. 4.11 the raw data of such a measurement is depicted in linear and logarithmic scale. One can easily distinguish the simultaneously measured two regimes. Moreover, one can see that the hologram is almost washed out due to the discussed poor resolution resulting from the large pinhole diameter compared to the cell size.

The idea behind the classification[22] scheme [27] now is that the measured diffraction patterns[23] are related to the spatial shape of the objects. In the case presented this is the dried cell, which in three dimensions might appear more like a pancake due to drying. At the nanoscopic scale cells and other biological specimens can have a rather

[21] For convenience the whole measured intensity distribution will be denoted as *diffraction pattern* in the remainder of this section, regardless of the fact that the center represents actually a hologram.

[22] A comparable Fourier transform based approach was proposed by Lendaris and Stanley for classifying photographic transparencies [29].

[23] Of course this method will also work with pure CDI setups and is not limited to the presented geometry, just the data collection then becomes more problematic due to the need for a beam stop etc.

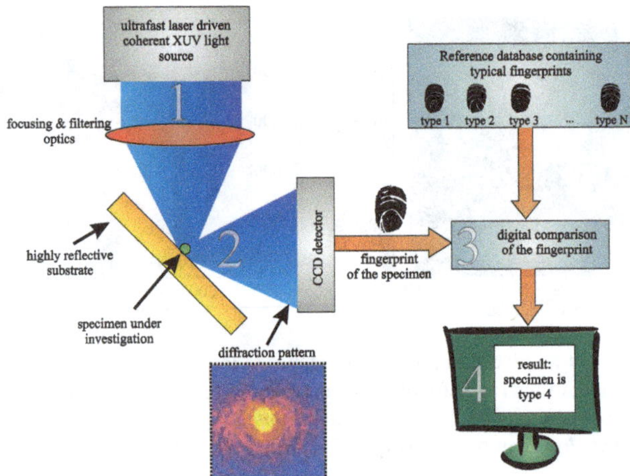

Fig. 4.12 General scheme for specimen classification using CDI techniques. An XUV light source (*1*) is spectrally and spatially filtered and then focused onto the specimen. The specimen is deposited onto a highly reflecting substrate in order to probe the silhouette and surface of the specimen at the same time. A CCD captures the generated diffraction pattern (*2*) in the far-field. This fingerprint of the specimen is digitally compared (*3*) with known fingerprints of the specimens under investigation from a database. The comparison results in the classification of the specimen (*4*). Using this scheme a rapid and high throughput classification even in a clinical environment might be possible

complex outer shape, see e.g. [1]. So, if one suspects that the same type of cells exhibit similar outer structures one should be able to identify the cells by their diffraction pattern. Hence, the need for an object plane image reconstruction is dispensed with and one can directly compare the diffraction patterns obtained from the cells. Since the intensity measurement in Fourier transform plane is insensitive to the phase it is also insensitive regarding the actual sample placement relative to the beam and the global phase shifts (see discussion in Sect. 2.2.3). This fact alleviates the classification in Fourier transform plane even further and one can expect that similar outer shapes of the cells will give rise to similar features in the diffraction pattern, allowing to distinguish them. Hence, for a clinical implementation, one would need a database with known diffraction patterns of the classes of objects one wants to identify, e.g. comparable to a database of fingerprints, and then use well-established and fast picture comparison techniques to determine the object. This process is illustrated in Fig. 4.12. This could in principle result in a high throughput technique provided a higher photon flux from the source can be achieved.

For practically testing this scheme the diffraction patterns of three single MCF7 cells and one SKBR3 cell were separately captured. Each of the cells was located inside another droplet, hence it dried separately and arranged itself independently from the other cells. The exposure time on each cell was $t_{\text{exp}} = 1{,}500\,\text{s}$ and two by two pixels on the CCD were binned into one. The exposures on each cell were repeated several times, which produced identical diffraction patterns. This justifies the fact that radiation damage is not limiting this experiment, in contrast to those

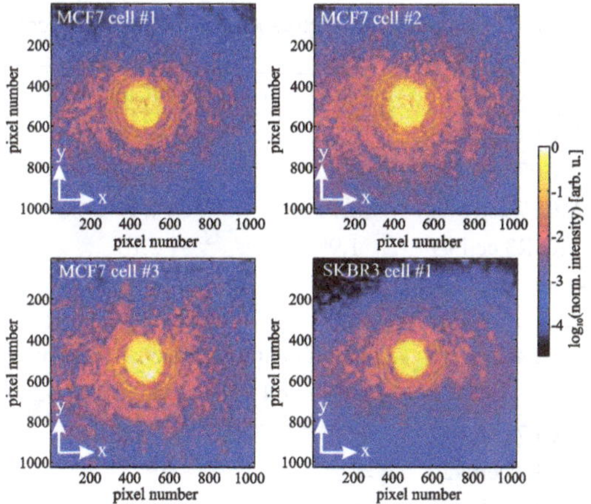

Fig. 4.13 Measured raw diffraction patterns of different breast cancer cells. Each panel shows a captured raw diffraction pattern of single different MCF7 and SKBR3 cells as indicated in each panel. All panels have the depicted logarithmic *color scale*. Despite each diffraction pattern is different one can already see with the bare eye that there is a difference between the two cell types

performed at free-electron lasers [12]. The four measured raw diffraction patterns are depicted in Fig. 4.13. Even with the naked eye one sees that the MCF7 patterns appear alike and the SKBR3 diffraction pattern differs.

In order to quantify this, first a 2D cross-correlation $C(i, j)$ is computed

$$C(i, j) = \sum_{m=0}^{N-1} \sum_{n=0}^{N-1} A(m, n) B^*(m + i, n + j) \tag{4.2}$$

of the images[24] $A(m, n)$ and $B(m, n)$, where m and n denote the pixel numbers in the original images and $0 \leq i, j \leq 2N - 1$ denotes the pixel coordinates in the cross-correlation image. In the next step a correlation factor κ is computed

$$\kappa = \max |\nabla[\nabla \mathrm{C(i, j)}]| \,, \tag{4.3}$$

where $\nabla C(i, j) = \frac{\partial C(i,j)}{\partial i}\hat{i} + \frac{\partial C(i,j)}{\partial j}\hat{j}$. The resulting κ gives an indication how well two images compare. For evaluating the data, κ is computed for each measured diffraction with one another. The result is depicted in Table 4.1. On the main diagonal high correlation factors for each diffraction pattern with itself are obvious. By introducing an experimentally determined threshold of $\kappa > 3.5$ in order to judge whether two diffraction patterns match or not, one can turn Table 4.1 into a cross validation matrix as shown in Table 4.2. The matrix demonstrates that this technique allows to distinguish between the cell types in this experiment. It is surprising that this works at all, because the cells are randomly oriented and the numerical aperture in this experiment (NA $\approx$ 0.04) does not allow for high resolution images.

[24] Here *image* refers to the measured raw data from the CCD. Each image was normalized to a common level.

Table 4.1 Correlations factors κ obtained from 2D cross-correlations of each diffraction pattern with one another

κ [arb. u.]	MCF7 cell #1	MCF7 cell #2	MCF7 cell #3	SKBR3 cell #1
MCF7 cell #1	13.03	9.48	3.75	1.99
MCF7 cell #2	9.48	15.52	4.49	2.16
MCF7 cell #3	3.75	4.49	13.67	3.13
SKBR3 cell #1	1.99	2.16	3.13	7.95

See text for details

Table 4.2 Cross validation matrix of the classification experiment

Class. result	MCF7 cell #1	MCF7 cell #2	MCF7 cell #3	SKBR3 cell #1
MCF7 cell #1	$=$	$=$	$=$	$\neq$
MCF7 cell #2	$=$	$=$	$=$	$\neq$
MCF7 cell #3	$=$	$=$	$=$	$\neq$
SKBR3 cell #1	$\neq$	$\neq$	$\neq$	$=$

This is the result from judging that diffraction patterns having a $\kappa > 3.5$ were produced from the same cell type. Please see text for further discussion

Of course the results presented in this chapter are far from a proof that this method would work with a thousand cells and hundreds of possible cell types. Moreover, no actual database exists yet. However, this work shows in a first experiment that there is a good indication that a diffraction pattern classification is possible. Much more precise results could be obtained by using more advanced image comparison techniques, such as those which are comprehensively outlined in [30]. This could allow for sampling patient tissue at a high throughput and low costs in a clinical environment, since the preparation of the samples is straightforward and cheap because labeling, staining or cultivating is not required. Pipetting the specimen onto a planar substrate is much easier compared to the deposition onto thin membranes, as outlined in Sect. 4.1.1. Other possible applications of this technique can be found in micromachining and the semiconductor industry, where e.g. a defective part could be simply identified by a deviating diffraction pattern. Moreover, the setup can be combined with other techniques that for instance analyze visible fluorescence emission triggered by the highly energetic photons of the XUV light.

In this section radiation damage effects to the biological specimens were completely neglected. This is a topic that is already extensively covered in literature [4, 31, 32].

Using the formulas given in [33] and the experimental parameters[25] the total dose can be estimated to $D \approx 7 \times 10^9\,\mathrm{Gy}$[26] which is far from threatening destruction to the sample [33]. Moreover, repeated measurements on the same cell resulted in

[25] The measured power of a single harmonic at 1 kHz repetition rate behind the pinhole is $\approx$1 nW. Together with $\lambda = 39\,\mathrm{nm}$ this yields a number of photons per pulse on target $n_{ph} \approx 6 \times 10^6$ photons/pulse. Together with the exposure time $t_{exp} = 1{,}500\mathrm{s}$ and the values and formulas given in [33] it allows an estimate of the dose.

[26] Unit: 1 [Gy] = 1 [J/kg] is the unit of the dose called *Gray*.

comparable diffraction patterns, hence no relevant radiation damage occurred. To further enhance the outlined method it would be beneficial to operate in the water window in order to acquire information from the organelles of the cell and thus enhance the classification of the specimen.

References

1. Jiang, H., Song, C., Chen, C.C., Xu, R., Raines, K.S., Fahimian, B.P., Lu, C.H., Lee, T.K., Nakashima, A., Urano, J., Ishikawa, T., Tamanoi, F., Miao, J.: Quantitative 3D imaging of whole, unstained cells by using X-ray diffraction microscopy. Proc. Natl. Acad. Sci. USA **107**(25), 11234–11239 (2010)
2. Boutet, S., Lomb, L., Williams, G.J., Barends, T.R.M., Aquila, A., Doak, R.B., Weierstall, U., DePonte, D.P., Steinbrener, J., Shoeman, R.L., Messerschmidt, M., Barty, A., White, T.A., Kassemeyer, S., Kirian, R.A., Seibert, M.M., Montanez, P.A., Kenney, C., Herbst, R., Hart, P., Pines, J., Haller, G., Gruner, S.M., Philipp, H.T., Tate, M.W., Hromalik, M., Koerner, L.J., van Bakel, N., Morse, J., Ghonsalves, W., Arnlund, D., Bogan, M.J., Caleman, C., Fromme, R., Hampton, C.Y., Hunter, M.S., Johansson, L.C., Katona, G., Kupitz, C., Liang, M.N., Martin, A.V., Nass, K., Redecke, L., Stellato, F., Timneanu, N., Wang, D.J., Zatsepin, N.A., Schafer, D., Defever, J., Neutze, R., Fromme, P., Spence, J.C.H., Chapman, H.N., Schlichting, I.: High-resolution protein structure determination by serial femtosecond crystallography. Science **337**(6092), 362–364 (2012)
3. Dierolf, M., Menzel, A., Thibault, P., Schneider, P., Kewish, C.M., Wepf, R., Bunk, O., Pfeiffer, F.: Ptychographic X-ray computed tomography at the nanoscale. Nature **467**(7314), 436–439 (2010)
4. Seibert, M.M., Boutet, S., Svenda, M., Ekeberg, T., Maia, F.R.N.C., Bogan, M.J., Timneanu, N., Barty, A., Hau-Riege, S., Caleman, C., Frank, M., Benner, H., Lee, J.Y., Marchesini, S., Shaevitz, J.W., Fletcher, D.A., Bajt, S., Andersson, I., Chapman, H.N., Hajdu, J.: Femtosecond diffractive imaging of biological cells. J. Phys. B At. Mol. Opt. **43**(19), 194015 (2010)
5. Henke, B.L., Gullikson, E.M., Davis, J.C: X-ray interactions: photoabsorption, scattering, transmission, and reflection at E=50-30000 eV, Z=1-92. At. Data Nucl Data. Tables **54**(2), 181–342 (1993)
6. Hardy, S.: Human Microbiology. Taylor and Francis, London (2002)
7. Plidschun, M.: Herstellung von nicht-periodischen Nanostrukturproben fuer XUV-Mikroskopie. Bachelor thesis, Friedrich-Schiller-University Jena (2013)
8. Young, K. D.: The selective value of bacterial shape. Microbiol. Mol. Biol. Rev. **70**(3), 660–703 (2006)
9. Seaberg, M.D., Adams, D.E., Townsend, E.L., Raymondson, D.A., Schlotter, W.F., Liu, Y.W., Menoni, C.S., Rong, L., Chen, C.C., Miao, J.W., Kapteyn, H.C., Murnane, M.M.: Ultrahigh 22 nm resolution coherent diffractive imaging using a desktop 13 nm high harmonic source. Opt. Express **19**(23), 22470–22479 (2011)
10. Szameit, A., Shechtman, Y., Osherovich, E., Bullkich, E., Sidorenko, P., Dana, H., Steiner, S., Kley, E.B., Gazit, S., Cohen-Hyams, T., Shoham, S., Zibulevsky, M., Yavneh, I., Eldar, Y.C., Cohen, O., Segev, M.: Sparsity-based single-shot subwavelength coherent diffractive imaging. Nat. Mater. **11**(5), 455–459 (2012)
11. Sandberg, R.L., Song, C., Wachulak, P.W., Raymondson, D.A., Paul, A., Amirbekian, B., Lee, E., Sakdinawat, A.E., La, O., Vorakiat C., Marconi, M.C., Menoni, C.S., Murnane, M.M., Rocca, J.J., Kapteyn, H.C., Miao, J.: High numerical aperture tabletop soft x-ray diffraction microscopy with 70-nm resolution. Proc. Natl. Acad. Sci. USA **105**(1), 24–27 (2008)
12. Chapman, H.N., Barty, A., Bogan, M.J., Boutet, S., Frank, M., Hau-Riege, S.P., Marchesini, S., Woods, B.W., Bajt, S., Benner, H., London, R.A., Plonjes, E., Kuhlmann, M., Treusch, R.,

Dusterer, S., Tschentscher, T., Schneider, J.R., Spiller, E., Moller, T., Bostedt, C., Hoener, M., Shapiro, D.A., Hodgson, K.O., Van der Spoel, D., Burmeister, F., Bergh, M., Caleman, C., Huldt, G., Seibert, M.M., Maia, F.R.N.C., Lee, R.W., Szoke, A., Timneanu, N., Hajdu, J.: Femtosecond diffractive imaging with a soft-X-ray free-electron laser. Nat. Phys. **2**(12), 839–843 (2006)
13. Raines, K.S., Salha, S., Sandberg, R.L., Jiang, H., Rodriguez, J.A., Fahimian, B.P., Kapteyn, H.C., Du, J., Miao, J.: Three-dimensional structure determination from a single view. Nature **463**(7278), 214–217 (2010)
14. Gardner, D.F., Zhang, B., Seaberg, M.D., Martin, L.S., Adams, D.E., Salmassi, F., Gullikson, E., Kapteyn, H., Murnane, M.: High numerical aperture reflection mode coherent diffraction microscopy using off-axis apertured illumination. Opt. Express **20**(17), 19050–19059 (2012)
15. Chen, C.C., Jiang, H.D., Rong, L., Salha, S., Xu, R., Mason, T.G., Miao, J.W.: Three-dimensional imaging of a phase object from a single sample orientation using an optical laser. Phys. Rev. B **84**(22), 224104 (2011)
16. Fienup, J.R., Crimmins, T.R., Holsztynski, W.: Reconstruction of the support of an object from the support of its auto-correlation. J. Opt. Soc. Am. **72**(5), 610–624 (1982)
17. Marchesini, S., He, H., Chapman, H.N., Hau-Riege, S.P., Noy, A., Howells, M.R., Weierstall, U., Spence, J.C.H.: X-ray image reconstruction from a diffraction pattern alone. Phys. Rev. B **68**(14), 140101 (2003)
18. Chapman, H.N., Barty, A., Marchesini, S., Noy, A., Hau-Riege, S.P., Cui, C., Howells, M.R., Rosen, R., He, H., Spence, J.C., Weierstall, U., Beetz, T., Jacobsen, C., Shapiro, D.: High-resolution ab initio three-dimensional x-ray diffraction microscopy. J. Opt. Soc. Am. A Opt. Image Sci. Vis. **23**(5), 1179–1200 (2006)
19. Zuerch, M., Kern, C., Spielmann, C.: XUV coherent diffraction imaging in reflection geometry with low numerical aperture. Opt. Express **21**(18), 21131–21147 (2013)
20. Martin, A.V., Wang, F., Loh, N.D., Ekeberg, T., Maia, F.R.N.C., Hantke, M., van der Schot, G., Hampton, C.Y., Sierra, R.G., Aquila, A., Bajt, S., Barthelmess, M., Bostedt, C., Bozek, J.D., Coppola, N., Epp, S.W., Erk, B., Fleckenstein, H., Foucar, L., Frank, M., Graafsma, H., Gumprecht, L., Hartmann, A., Hartmann, R., Hauser, G., Hirsemann, H., Holl, P., Kassemeyer, S., Kimmel, N., Liang, M., Lomb, L., Marchesini, S., Nass, K., Pedersoli, E., Reich, C., Rolles, D., Rudek, B., Rudenko, A., Schulz, J., Shoeman, R.L., Soltau, H., Starodub, D., Steinbrener, J., Stellato, F., Strueder, L., Ullrich, J., Weidenspointner, G., White, T.A., Wunderer, C.B., Barty, A., Schlichting, I., Bogan, M.J., Chapman, H.N.: Noise-robust coherent diffractive imaging with a single diffraction pattern. Opt. Express **20**(15), 16650–16661 (2012)
21. Sun, T., Jiang, Z., Strzalka, J., Ocola, L., Wang, J.: Three-dimensional coherent X-ray surface scattering imaging near total external reflection. Nat. Photon. **6**(9), 588–592 (2012)
22. Weise, F., Neumark, D.M., Leone, S.R., Gessner, O.: Differential near-edge coherent diffractive imaging using a femtosecond high-harmonic XUV light source. Opt. Express **20**(24), 26167–26175 (2012)
23. Tripathi, A., Mohanty, J., Dietze, S.H., Shpyrko, O.G., Shipton, E., Fullerton, E.E., Kim, S.S., McNulty, I.: Dichroic coherent diffractive imaging. Proc. Natl. Acad. Sci. USA **108**(33), 13393–13398 (2011)
24. McClatchey, K.D.: Clinical Laboratory Medicine. Lippincott Williams and Wilkins, Philadelphia (2002)
25. Rosch, P., Harz, M., Schmitt, M., Peschke, K.D., Ronneberger, O., Burkhardt, H., Motzkus, H.W., Lankers, M., Hofer, S., Thiele, H., Popp, J.: Chemotaxonomic identification of single bacteria by micro-Raman spectroscopy: application to clean-room-relevant biological contaminations. Appl. Environ. Microbiol. **71**(3), 1626–1637 (2005)
26. Ashton, L., Lau, K., Winder, C.L., Goodacre, R.: Raman spectroscopy: lighting up the future of microbial identification. Future Microbiol. **6**(9), 991–997 (2011)
27. Zuerch, M., Spielmann, C.: Patent: Verfahren zur Auswertung von durch schmalbandige, kurzwellige, kohaerente Laserstrahlung erzeugten Streubildern von Objekten, insbesondere zur Verwendung in der XUV-Mikroskopie. DE102012022966.6 - int. reg. PCT/DE2013/000696, 21 Nov 2012 (2012)

28. Williams, G.J., Quiney, H.M., Dhal, B.B., Tran, C.Q., Nugent, K.A., Peele, A.G., Paterson, D., de Jonge, M.D.: Fresnel coherent diffractive imaging. Phys. Rev. Lett. **97**(2), 025506 (2006)
29. Lendaris, G.G., Stanley, G.L.: Diffraction-pattern sampling for automatic pattern recognition. Pr. Inst. Electr. Elect. **58**(2), 198–216 (1970)
30. Saleh, B.: Introduction to Subsurface Imaging. Cambridge University Press, New York (2011)
31. Mancuso, A.P., Gorniak, T., Staler, F., Yefanov, O.M., Barth, R., Christophis, C., Reime, B., Gulden, J., Singer, A., Pettit, M.E., Nisius, T., Wilhein, T., Gutt, C., Grubel, G., Guerassimova, N., Treusch, R., Feldhaus, J., Eisebitt, S., Weckert, E., Grunze, M., Rosenhahn, A., Vartanyants, I.A.: Coherent imaging of biological samples with femtosecond pulses at the free-electron laser FLASH. New J. Phys. **12**, 035003 (2010)
32. Garman, E.F., McSweeney, S.M.: Progress in research into radiation damage in cryo-cooled macromolecular crystals. J. Synchrotron. Radiat. **14**(Pt 1), 1–3 (2007)
33. Howells, M.R., Beetz, T., Chapman, H.N., Cui, C., Holton, J.M., Jacobsen, C.J., Kirz, J., Lima, E., Marchesini, S., Miao, H., Sayre, D., Shapiro, D.A., Spence, J.C., Starodub, D.: An assessment of the resolution limitation due to radiation-damage in x-ray diffraction microscopy. J. Electron Spectros. Relat. Phenom. **170**(1–3), 4–12 (2009)

Chapter 5
Optical Vortices in the XUV

5.1 Singular Light Beams—Optical Vortices

In the preceding chapters and almost everywhere in imaging theory, flat phases and plane waves are assumed for illumination. This is convenient, because it yields a simple solution of the Helmholtz equation and thus obeys Maxwell's equations. However, there is a manifold of possible known other solutions. One of these quite special solutions are so-called optical vortex (OV) or singular light beams. The peculiarity of these beams is that they have a radial screw-like phase distribution. This characteristic helical phase profile can be described by $\exp(-il\phi)$, where l is an integer number and ϕ is the azimuthal coordinate. For a physical solution the radial phase must be wrapped modulo 2π along the azimuthal coordinate. This leads to a phase singularity in the center of such beams, where the phase is undefined. In turn the intensity at this phase singularity becomes zero, because the real and imaginary part of the field amplitude vanish. This can be also interpreted as destructive interference, since all phases are present at the same time at the singularity with equal amplitude [1]. Such screw-type distributions with a physical quantity becoming zero in the center are well-known from nature and compare e.g. to a water vortex, hence the name *optical vortex*. The number l is also called the *winding number* or the *topological charge* (TC). The first actual description of such singularities in optics goes back to a work from Nye and Berry [2]. Nowadays, optical vortices are a large field of research, a comprehensive overview can for instance be found in [3].

This thesis will thus focus on a few relevant aspects and introduce OV phenomenologically. For this, a closer look is taken at the field amplitude $u_p^l(r, \phi, z)$ of such beams. Its radial dependence $u_p^l(r, \phi, z)$ can be expressed by Laguerre-Gaussian (LG) modes [4], described by

M.W. Zürch, *High-Resolution Extreme Ultraviolet Microscopy*, Springer Theses, DOI 10.1007/978-3-319-12388-2_5

$$u_p^l(r,\phi,z) = \frac{C}{(1+z^2/z_r^2)^{\frac{1}{2}}} \left(\frac{r\sqrt{2}}{w(z)}\right)^l L_p^l \left(\frac{2r^2}{w^2(z)}\right) \exp\left(-\frac{r^2}{w^2(z)}\right)$$
$$\exp\left(-\frac{ikr^2 z}{2(z^2+z_r^2)}\right) \exp(-il\phi) \exp\left(i(2p+l+1)\tan^{-1}\frac{z}{z_r}\right), \tag{5.1}$$

which are eigenmodes of the paraxial Helmholtz equation. In this equation C is a constant, k is the wavevector, z_r is the Rayleigh length, and L_p^l are the generalized Laguerre polynomials with radial index p and a beam radius $w(z)$ (beam waist at $z = 0$) known from Gaussian optics. The LG modes with $p = 0$ feature a bright ring with a radius $R(Z) \propto w(z)\sqrt{l}$, which obviously increases with the TC [5].[1] This leads to the possibility of trapping particles in a tight focus of an OV beam, because it forms a toroidal trap where the particles are repelled from the high intensity ring [6, 7]. Another property of OV beams can be found by inspecting the Poynting vector $\mathbf{S} = \mathbf{E} \times \mathbf{H}$, which describes the momentum flow in free space. For the screw-type phase profile the Poynting vector has an azimuthal component, leading to a Poynting vector that describes a spiral along the propagation direction [4]. This results in an orbital angular momentum (OAM) parallel to the beam axis, because the Poynting vector describes the energy transfer in free space. It was found by semiclassical calculations that each photon in a such a helical beam carries an OAM of $l\hbar$ [8] independent of its spin, i.e. its polarization state. Hence, such photons can transfer a torque on matter upon interaction. This was already postulated by POYNTING himself for circular light and experimentally measured for the first time by Beth [9]. Hence, such beams can be used to transfer momentum to particles [1, 10] or even micromachines [7, 11]. The implications from the OAM are, however, more far-reaching, because in optics typically every photon that is absorbed or emitted from an atom exchanges a quanta $\pm\hbar$ due its spin. However, if a photon additionally carries an OAM an additional degree of freedom arises, which allows to address transitions that are forbidden by the selection rules in the electric and magnetic dipole approximation [12]. Other potential applications can for instance be found in quantum information [13] and interferometry [14].

There are various techniques to generate an OV. The most obvious one is to use a radial phase plate [15] or vortex lenses [16]. The difficulty with those is that the radially etched steps must be accurately matched to the wavelength and typically the phase profile must be discretized for practical fabrication. Other techniques include astigmatic laser mode converters [17] and holographic phase imprint [18]. However, special care has to be taken if dealing with ultrafast light that is to be converted, because the large bandwidth makes the proper phase shaping experimentally difficult. Special dispersion-compensating setups were invented for such purposes [19]. In this work, however, a spatial light modulator (SLM) was used to imprint the phase onto a Gaussian input beam. Using SLMs to generate OV beams was successfully

[1] Experimentally a radius increase $R \propto l$ was found in [1].

demonstrated before [1]. The technical realization done in this work will be described in full depth in the next section.

The objective of dealing with OV in this work was to investigate whether such OV could be transfered into the XUV regime by HHG (Sect. 2.1). The reason as to why it is important to generate OVs at shorter wavelengths is that the spectroscopy of forbidden inner transitions in atoms [12] requires typically higher photon energies than would be achievable in the visible light domain. A harmonics comb produced with HHG would thus be perfect for spectroscopy if the OAM could be transferred to each individual harmonic. The general feasibility of generating OVs at soft X-ray wavelengths was demonstrated using synchrotron radiation and an appropriate phase plate [20, 21]. To access even higher topologic charges at X-ray wavelength, high precision phase plates were successfully tested [22]. For the wavelengths produced by HHG, however, phase plates are not an option due to the high absorption of matter.

Unfortunately, such OV beams do not behave in an orderly manner in nonlinear light-matter interactions. The vortices tend to break-up in self-focusing nonlinear media. This was first predicted in [23] and observed in [24, 25]. This is also known as *optical vortex decay*, which eventually results in a splitting of a l-fold screw dislocation into l dislocations with unit charge in the presence of a small coherent pedestal [26]. This was experimentally observed e.g. in [27, 28]. Moreover, modulation instabilities, i.e. small imperfections in the OV beam profile, can lead to the same effect in the presence of a nonlinearity [29].

5.2 Experimental Setup for Generating XUV Vortex Beams

For the OV experiments presented in this work a spatial light modulator (SLM) was used to imprint the typical screw-like phase profile (Fig. 5.1b) onto a Gaussian near-infrared beam used to produce the harmonics. This principle is well-known

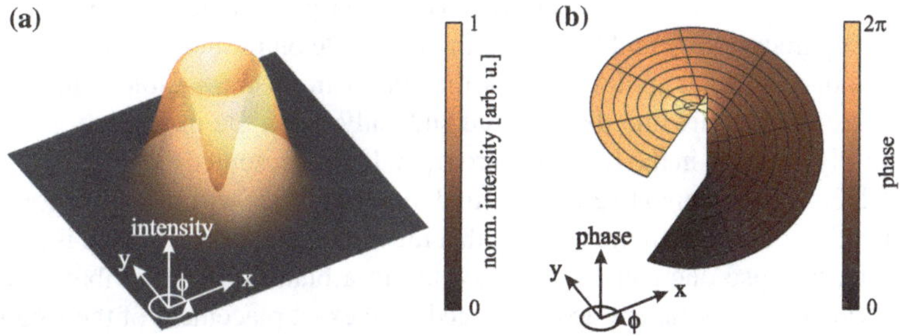

Fig. 5.1 Principal structure of an optical vortex. **a** The intensity distribution features a *bright ring* and a *dark* core in the *center* of the vortex. **b** The phase distribution resembles a screw from zero to 2π for the presented OV having a TC of one. The phase singularity in (**b**) coincides with the *center* of the *dark* core in (**a**). OVs with higher TCs exhibit multiple phase wraps modulo 2π in one turn around the azimuthal angle ϕ. In that case the diameter of the *bright ring* in (**a**) increases as well according to the TC

and has been used to generate OV beams before [1]. The SLM (type HAMAMATSU LCOS- SLM X10468) consists of a highly reflecting mirror with a thin nematic liquid crystal array on top. Areas of 20 μm by 20 μm can be addressed like a pixel on a computer display, making up a total array of 800 by 600 pixels. In order to modulate a beam it is simply reflected off this device. The corresponding part of the beam that is reflected off a pixel experiences a refractive index change corresponding to the state of the nematic liquid crystal in that particular pixel. Each pixel has a control depth of 8 bits. The specific device used in this work allows a phase shift per pixel of a little more than 2π for wavelengths 800 ± 50 nm. The device is addressed by the graphics card of a common desktop computer using gray-scale images. A black colored pixel, i.e. pixel value zero, and a white pixel, i.e. pixel value 255, correspond to zero and $\approx 2.2\pi$ phase shift respectively. A global phase-shift is added due to the light passing through the crystal, which is the same for all pixels. The total thickness of the liquid crystal layer and the cover glass is $\sim$1 mm, hence dispersion can be neglected for the 30 fs pulses used in the experiment presented. In summary, the SLM device described above is readily capable of modulating an impinging beam to imprint a helical phase profile in order to generate an OV beam.[2]

The SLM was placed in the Ti:Sa driven HHG setup (Fig. 3.2a) before the final focusing optic into the HHG target. A sketch of the whole setup that is discussed in this and the following paragraphs is depicted in Fig. 5.2a. The angle of incidence on the SLM was kept as low as possible. A self-programmed LabView program generates the gray-scale image that is subsequently transfered onto the SLM. The program allows to change the topological charge, the gray-value for 2π and the actual position of the singularity with respect to the image coordinates. The latter proved very important, because the singularity must be placed as accurately as possible on the center of the impinging Gaussian beam. If the singularity was placed more than $\sim$10 pixels off the proper center it was not possible to observe any vortex structure at all in the XUV. The size of the impinging laser beam was adapted to the overall size of the active SLM area by a telescope. In the experiments the singularity was first roughly placed on the Gaussian mode by visual inspection on a screen behind the SLM, the dark vortex core was clearly discernible. The exact gray-value that corresponds to 2π was experimentally found by inspecting the mode on the screen. If the value did not correspond to 2π an additional dark line with fringes was visible at the position of the phase jump. This line disappeared suddenly when the maximum gray-value was carefully adjusted in the LabView program. In the presented experiments a gray-value of 233 corresponding to 2π was found. It had to be set within ± 1 bits for proper OV generation.[3] It is worth mentioning that the SLM generation method is also very convenient, because one can instantly switch to a blank phase distribution on the SLM leaving the Gaussian mode unchanged. The exact placement of the singularity

[2] Higher topological charges can be generated by introducing several 2π phase wraps around the azimuthal coordinate.

[3] Actually, a high TC was set first, which resulted in plenty of lines at every phase wrap position looking like a wagon wheel on the screen if the gray-value was not properly set. The alignment of the correct gray-value was straightforward that way.

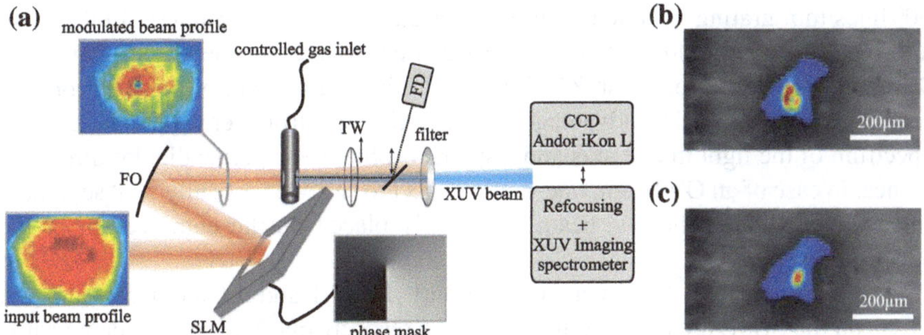

Fig. 5.2 Principal setup of the optical vortex experiment. **a** The input laser beam, having almost an Gaussian shape, from a femtosecond Ti:Sa laser system is modulated by a spatial light modulator (SLM), which is fed with a *gray-scale* image containing the helical phase mask. The two *insets* show the measured beam profiles before and after the SLM. The modulated beam clearly features the *bright ring* and *dark* core of an optical vortex. A focusing optic (FO), in the experiment a $f = 300\,\text{mm}$ lens, focuses the modulated beam into the HHG setup described in Sect. 3.1.1. Several diagnostics can be placed behind the HHG source. A $5\,\mu\text{m}$ tungsten wire (TW) affixed to a frame can be inserted 80 mm behind the source as a wavefront splitting interferometer. A pickup mirror allows to feed the infrared laser pulses into a focus diagnostic setup (FD). A total of $0.5\,\mu\text{m}$ thick aluminum is employed to filter the infrared light. The pure XUV beam can either be measured spatially in the far-field by a CCD or spectrally measured after refocusing into a XUV imaging spectrometer. The measured laser focus using the FD with a phase profile for $\text{TC} = 1$ and a flat phase applied to the SLM are depicted in (**b**) and (**c**), respectively. Both pictures were overlayed with a properly scaled microscopic image of the hole drilled into the nickel tube in order to demonstrate that both modes fit well through the interaction region

on the Gaussian input mode was later determined in the experiment by analyzing the HHG signal.

Subsequently, the modulated beam was focused down into the HHG setup, which was previously described in Sect. 3.1.1. The intensity profile of an OV beam can described by Laguerre-Gaussian modes as described in the previous section. The intensity profile of an OV with a TC of one can be described by a LG_0^1 mode, while a standard Gaussian beam would be represented by LG_0^0. Hence the doughnut-like intensity distribution of the OV in the focus will be significantly larger than the Gaussian focus. Thus before the experiment the focal lens was translated relative to the nickel tube with a Gaussian beam applied. This results in drilling the hole bigger, such that the OV beam can also fit the interaction region (see overlay in Fig. 5.2b, c). A focus diagnostic could be employed by driving a pick-up mirror into the beam path behind the nozzle. For simple phase measurements by means of a wavefront splitting interferometer a $5\,\mu\text{m}$ thin tungsten wire could be placed 80 mm behind the source. After an aluminum filter ($0.5\,\mu\text{m}$ total thickness) an XUV sensitive CCD could be placed downstream for spatial inspection of the far-field. The total distance between source and CCD was approximately 600 mm. A Hartmann mask could be placed in front of the CCD to evaluate the wavefront. For spectral characterization an XUV imaging spectrometer (McPherson 248/310G with MCP imaging unit,

300 lines/mm grating installed) could be placed instead of the camera. In that case a toroidal mirror should be placed in between the source and the spectrometer in a $4f$-setup in order to focus the XUV light onto the entrance slit of the spectrometer (compare to Fig. 3.4). The advantage of an imaging spectrometer is that one gets the spectrum of the light in one axis and a spatial profile in the perpendicular direction. Hence, in case of an OV beam, one would expect to see two vertically well separated intensity maxima if the OV beam is properly placed on the entrance slit of the spectrometer.

The measured foci for a vortex beam with TC$=$ 1 and a Gaussian beam (flat phase on the SLM) are depicted in Fig. 5.2b, c respectively. The vortex beam profile is obviously not transfered perfectly into the focus. The missing dark vortex core in Fig. 5.2b could, however, be subject to the limited resolution and dynamic of the focus diagnostics. The asymmetry of the focus is likely caused by astigmatism, which is a known problem of the laser setup used. The astigmatism is mainly induced in an off-axis telescope that is used to adjust the beam size from the output of the Ti:Sa laser system to the SLM's active area.

5.3 First Demonstration of Optical Vortices in the XUV

When the helical phase profile corresponding to TC$=$ 1 was properly imprinted on the fundamental infrared beam it was possible to observe a structure similar to an optical vortex in the XUV. In order to suppress any mixing effects on the CCD by multiple harmonics being detected at the same time, the laser energy, and thus the intensity in the focus and thereby the cut-off, was reduced. Only the 11th harmonic at $\lambda = 74$ nm could proceed through the Al filter.[4] The resulting XUV light was spatially detected in the divergent beam behind the filter at a distance of 600 mm, see Fig. 5.3a. One finds the principal structure of an OV, i.e. a bright ring-like structure and a dark core indicated by the dotted line and the cross, respectively. The slight background could be incoherent emission from the source (compare to the background in the spectrum in Fig. 3.2b). If one now switches the SLM to a blank screen, one can instantly propagate a Gaussian beam through the system and generate an almost Gaussian-shaped harmonic beam (Fig. 5.3b). The center of the Gaussian XUV beam corresponds exactly to the position of the almost dark core in the XUV OV beam (cross in Fig. 5.3a). One should note that the pulse energy for the unmodulated beam was further reduced, such that only one harmonic could proceed through the Al filter. The imperfect shape of the OV XUV beam is explained by the imperfect shape of the generating beam in the focus (Fig. 5.2b). Furthermore, phase matching effects could play a role [31] in the way that the additional phase of the wavefront could cancel the HHG phase matching achieved by the Gouy-phase for certain azimuthal angles.

[4] At 74 nm half a micron Al has a transmission of $\sim$6 %. The 9th harmonic is totally suppressed already [30].

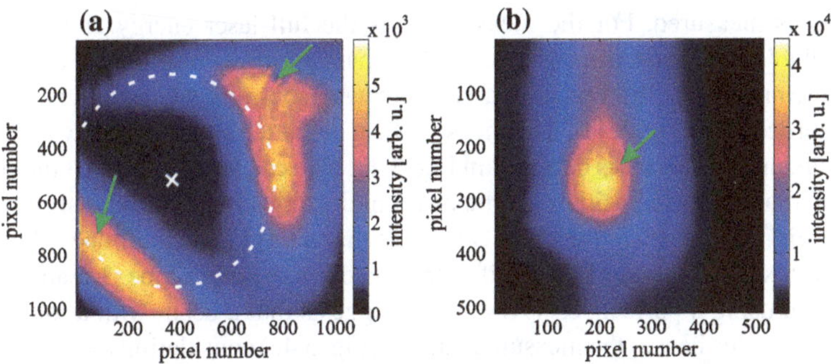

Fig. 5.3 Measured XUV beam at 74 nm with a TC = 1 optical vortex mode and Gaussian mode in the fundamental. **a** Seeding an HHG experiment with a TC = 1 infrared beam yields a doughnut-like intensity distribution that, apart from some distortions, resembles an optical vortex. The *dotted line* guides the eye on a *circle*. The center features an almost *dark* core, with only a few hundred counts remaining above the noise level, which are likely incoherent background. The center, marked by a *cross*, is exactly the position where a Gaussian mode (**b**) appears, if the SLM is switched to a blank screen. The latter corresponds also to a Gaussian fundamental mode. Compare to the measured corresponding focal beam profiles in Fig. 5.2b, c. The exposure time was $t_{exp} = 20\,\mathrm{s}$ and $t_{exp} = 0.2\,\mathrm{s}$ for the vortex and Gaussian fundamental mode, respectively. Two by two pixels on the CCD were binned into one superpixel in (**b**). The CCD was cooled to $-10\,^{\circ}\mathrm{C}$ and the electronic background was properly subtracted for each measurement. The selected frame of the CCD is identical in (**a**) and (**b**) and corresponds to 1024 by 1024 pixels on the CCD. The *vertical* smearing in (**b**) is an artifact from the camera readout. The distortions, examples marked with the *green arrows*, come from debris on the Al filter

The spatial shape alone, however, is no proof for the existence of an OV. Ideally the amount of topological charge carried in an OV beam is measured by interfering the OV beam with itself [32] or a Gaussian reference wave [33], which then readily proves the existence of an OV. Such measurements are straightforward in the visible light domain but remain difficult in the XUV and soft X-ray range, because 50/50 beam splitters are difficult to implement. Interference with a Gaussian reference could in principle be implemented more easily by splitting the fundamental infrared beam and build two independent HHG sources. However, with the available laser system the pulse energy was not sufficient. Hence, the wavefront splitting method suggested by Peele and Nugent [21] was employed in this work.

In this method an opaque wire is placed in the beam and one observes interference fringes in the far-field behind the wire. The tilt and shape of the fringe pattern is dependent on the phase of the wavefront at positions across the wire. For an OV beam the wavefront is tilted to opposite directions on opposite sides of the beam due to the screw-like phase profile, thus the fringes are laterally shifted with respect to the fringes of the opposite side of the beam.[5] By comparing the measured fringe pattern with simulations one can at least roughly determine the TC. In this work only an imperfect vortex signature was found. However, it was possible to place a $5\,\mu\mathrm{m}$ tungsten wire across the beam such that it intersected the parts where

[5] Here a wire centered on the singularity is expected, as was the case in the experiment.

intensity is measured. For the measurements the full laser energy and hence the full available HHG spectrum was used. The main spectral contribution is expected around 36 nm (Fig. 3.2b). For comparison a fringe pattern for a Gaussian input mode was recorded first (Fig. 5.4a). The measured fringe pattern for a vortex with TC = 1 in the fundamental beam is depicted in Fig. 5.4b. One can already see that the fringes are shifted at opposite positions of the beam indicating a relative phase shift (dotted lines in Fig. 5.4b to guide the eye). One finds that a bright fringe in one lobe of the XUV vortex ends on a dark fringe if extrapolated to the other lobe. An appropriate simulation using a perfect XUV OV beam suggests that a beam with a topological charge of one fits best to the measured pattern (Fig. 5.4c) considering the width of the lobes and the fringe shift. The simulation parameters and simulations for higher TCs are presented in appendix B. Another experimental step was to measure the spectrum of the XUV vortex beam with an imaging spectrometer, see Sect. 5.2 for details. The result is depicted in Fig. 5.4d. First, one finds that in all measured harmonic orders an intensity structure is evident. The two intensity maxima on the spatial coordinate correspond to the two lobes of the intensity distribution (Fig. 5.3a) being cut out of the beam by the entrance slit of the spectrometer. Second, it is evident that the two maxima are separated approximately by the same distance for all harmonics. Speaking of optical vortices one could attribute this to all harmonics having the same

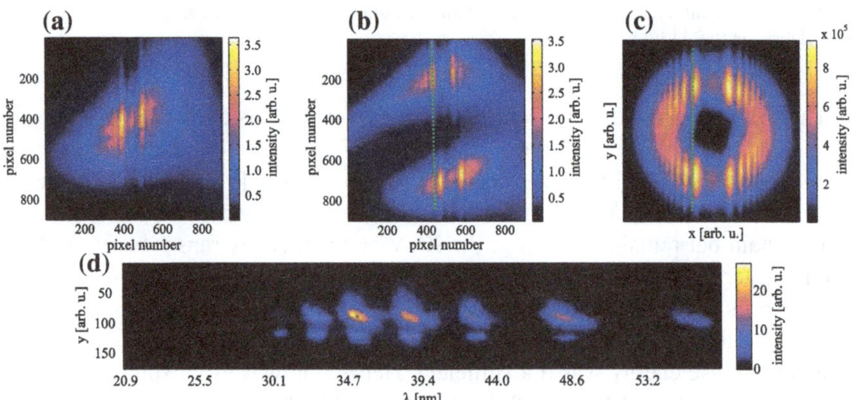

Fig. 5.4 Thin wire interference and spectrum of an XUV vortex beam. **a** The measured far-field fringe pattern caused by a 5 μm tungsten wire on a Gaussian shaped input pulse features parallel fringes. The full HHG spectrum behind the Al filter was used (compare to Fig. 3.2b). **b** If the vortex phase profile is added to the fundamental beam the characteristic structure (compare to Fig. 5.3a) indicating an XUV vortex beam appears. The fringes caused by the same wire are now shifted and one *bright* fringe in the *upper lobe* ends, if extrapolated, approximately at a *dark* fringe in the *lower lobe* (*dotted green line*). **c** Simulation of the fringe pattern caused by a TC = 1 vortex on a thin wire. See Appendix B for details and simulations for higher TCs. The width of the lobes in (**b**) and the fringe shift compare well. **d** Using an imaging spectrometer with a slightly opened entrance slit one finds that the XUV vortex signature is present in all visible harmonic orders. The experimental conditions compare to those for (**b**). Please note that the wire was slightly horizontally translated between (**a**) and (**b**) such that it covers the high intensity parts best

TC, because, as discussed in the fundamentals (Sect. 5.1), a higher TC would yield a larger ring diameter, which should in turn be visible by a larger separation. A further measurement was a wavefront analysis based on a Hartmann measurement with the implementation detailed in [34]. It resulted in the expected tilt of the wavefront, at least at the places of the beam where intensity was measured. The TC cannot be retrieved from that measurement in an accurate way. It was also tried to seed the HHG experiment with a TC$=2$ beam, which gave rise to a larger diameter of the XUV vortex beam. Unfortunately, the yield was extremely low, preventing a detailed analysis.[6] The results of this work were recently published [35].

More recently a theoretical work published by Hernández-García et al. [36] suggests, however, that a vortex beam having a m-fold topological charge in the fundamental should be $n \times m$ charged in the n-th harmonic in a frequency up-conversion experiment. This is intuitively clear if one considers a periodic process with a radiating dipole $d(t) = \mathrm{f}\left[\exp(\mathrm{i}[\omega t + \phi])\right]$, which is a function of the phase of the driving field. One can calculate the n-th harmonic of this field by a Fourier transform $\tilde{d}(n\omega) = \int \exp\left[in\omega\right] d(t)\mathrm{d}t$, which gives rise to the phase relation $\phi(n\omega) = n\phi(\omega)$. The spectral observation with an almost constant vortex radius beam across the harmonic orders was interpreted in [35] as an indication for a vortex beam with TC$=1$. If theoretically calculated as in [36] it turns out, however, that one would actually expect that for two adjacent harmonics n and $n+2$, with TCs behaving like $\mathrm{TC}_{n+2} = \mathrm{TC}_n + 2$, the radius of the ring would be almost equal, because a higher ring diameter due to the TC is compensated by the lower wavelength of the corresponding harmonic. Hence, the interpretation of the measured TC in the present experiment must be somewhat softened at this point. However, the fringe pattern behind the wire still prohibits the existence of a TC$=23$, as should be expected for the 23rd harmonic. Thus the final interpretation of this experiment still is that likely an optical vortex with a unit topological charge in the far-field was produced in the XUV by HHG. The explanation for this behavior is that directly in the interaction region vortices carrying high topological charges are generated. As discussed in the fundamentals, such highly charged vortices exhibit a high instability [37, 38] and can decay into several singly charged vortices which all carry the same sign. This behavior was also observed in second harmonic generation [28]. Those vortex cores then strongly repel each other. It eventually leads to the observation of a stable vortex that carries only a single topological charge in the far-field. The imperfect shape of the OV is explained by the imperfect focus of the infrared laser. The astigmatism on that beam may additionally increase the probability of the vortex decay according to the literature cited in the fundamentals.

What still remains unclear in that explanation is where the remaining 22 vortices disappear to. If they totally disappear in the far-field, they would need a large transversal k vector after the decay. This seems, however, not possible due to the fact that vortices are paraxial objects. In the experiment no evidence for the remaining 22

---

[6] Due to the increased diameter of the bright vortex ring in the fundamental the local intensity decreases for increasing topological charge, hence the HHG cut-off is significantly reduced for vortex beams carrying higher topological charges.

single charged vortices was found and, moreover, the observed structure was stable over longer periods, i.e. several minutes, which is not expected for a turbulent regime of vortices dynamically decaying and repelling one another.

Another reason for the observed intensity profile could be off-axis phase matching [39]. This was ruled out by scanning the gas backing pressure of the source, which significantly influences the general phase matching of the HHG radiation. If off-axis phase matching was an effect, then the diameter of the structure should be dependent on the gas backing pressure. The intensity signature in the XUV, however, remained the same with gas pressure, which rules out this effect. In any case, this could not cause the observed phase shift in the wavefront splitting experiment.

Although the explanation provided is to some degree unsatisfactory, it does facilitate a description of the features observed experimentally. However, there is no doubt that optical vortex beams can be produced in the XUV by direct conversion from the infrared by HHG. Further measurements on that topic should be done in the future to learn more about the generation process and how to improve it. One possible approach would be a clean-up of the Ti:Sa laser output by means of a spatial filter before the vortex signature is imprinted. Unfortunately, in the case presented this would have reduced the pulse energy too much for sufficient HHG, hence spatial cleaning would require a more powerful laser. Another option would be the use of a feedback algorithm to optimize the phase mask on the SLM, analogous to what was published for HHG enhancement by spatial beam shaping [40]. The idea would be to allow a small additional phase variation on the general vortex phase signature in order to account for aberrations that are already present on the input laser beam. This could be implemented by a feedback loop that gradually improves the phase mask in order to get a perfectly shaped vortex beam in the XUV.

5.4 Application of XUV Vortex Beams to High Resolution Lensless Imaging

In the previous section it was demonstrated that optical vortex beams can be generated in principle from an infrared vortex beam by HHG. In this section the theory will be described as to how these beams, provided they had a perfect shape, could enhance the achievable resolution in an imaging experiment.

The field of superresolution imaging is wide and draws on imaging well below the resolution limit given by the wavelength of the used light. Some of these techniques rely on measuring the evanescent field [41] or the optical near-field (NSOM) [42], which includes scanning of the sample. Other techniques operate in the far-field and use luminescence of the object itself (STED) [43], which also needs scanning of the sample. Another approach is structured illumination microscopy (SIM) [44]. All of these methods were demonstrated with visible light so far and applying those methods to XUV wavelengths could boost the achievable resolution even further. For imaging with optical vortices two principal methods arise. First, one could exploit the fact that the radial phase distribution on an optical vortex is continuous and hence samples interferences more finely than the wavelength. That is analogous to the discussion of resolving depth profiles of objects below one wavelength from longitudinal phase shifts. A first demonstration using the phase profile of OV beams for near-field control is already reported [45]. A resolution comparable to a quarter of the illumination wavelength was reported for imaging based on surface plasmon waves induced by OV beams [46]. In Fig. 5.5 a numerical simulation on that matter is presented. An 80 nm wide μ-shaped aperture is placed in the focus of an optical vortex beam that has a width of roughly two microns. In Fig. 5.5a–c the intensity of the illumination field is overlayed with the aperture to get an impression of the spatial relations. The wavelength was set to $\lambda = 38$ nm for all simulations. Fig. 5.5a shows a TC = 0 beam, i.e. a Gaussian-shaped beam, as reference. The corresponding far-field (Fig. 5.5d) features only a few smeared out fringes that could be detected. For the vortex beams with higher topological charges (Fig. 5.5b, c) the pattern features many more fringes, which become finer for increasing topological charge. Although not explicitly shown, one could certainly expect a better resolution in the reconstruction of these fields by an adapted HIO algorithm, where the radial phase distribution is assumed in the illumination field. This is mainly explained by the finer fringes, which from physical intuition should yield sharper details in the image.

The second, somewhat more extraordinary, possibility for imaging with OV beams would use the dark core of the singularity as a probe. As has been discussed in the fundamentals section and observed in the experiments, even small perturbations already lead to a break up of a highly charged vortex, as they should in principle appear in HHG, into several single charged vortices. These singularities are not subject to diffraction and could thus propagate into the far-field, where the corresponding positions of the dark cores can easily be measured by a CCD. Hence, it appears possible to acquire information on the nanoscopic interaction region of the vortex core by tracing the singularities in the far-field. This has been numerically demonstrated recently [47]. The idea of that publication is mainly that the vortex structure behind an interaction with matter depends on the structure of the medium. Hence, these singularities could well be used as a probe. By scanning the OV beam across the sample one could thus acquire a high-resolution image of the sample. The size of the singularity is by definition infinitesimally small and is independent of the wavelength. The size of the dark core is, however, related to the topological charge and the overall size of the beam (Eq. 5.1). Hence, the size of the probe is dependent on the focusability of the overall beam and thus the shorter wavelength could allow for much tighter focusing of the full beam. This would then allow for even higher resolution in OV imaging in the XUV compared to visible light. Moreover, another great benefit of this technique would be that there is zero intensity at and around the sample, hence, damaging the sample with the high intensity radiation would be no issue.

Unfortunately, it is not possible at this stage to give accurate numbers about the achievable resolution if a vortex beam is used for illumination, because nothing accurate, or even a proof-of-principle experiment, has been published about this topic yet. However, it would certainly be beneficial to conduct research on that, as

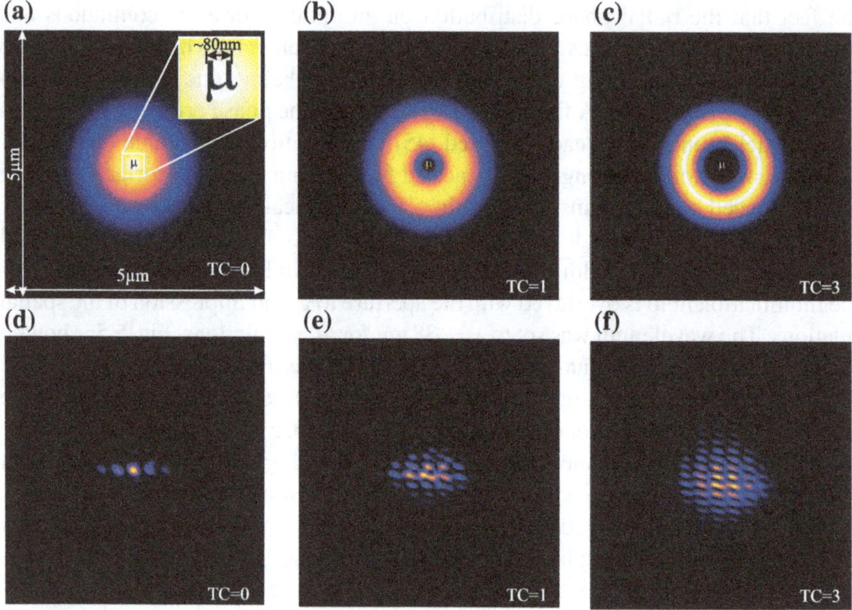

Fig. 5.5 Simulation of imaging an aperture with an OV beam. A μ-shaped aperture with a width of approximately 80 nm is illuminated with **a** TC = 0 (i.e. Gaussian shape), **b** TC = 1 and **c** TC = 3 optical vortex beam having a diameter of roughly two microns. In (**a**) the aperture (*inset*) is colored *black*, while it is colored *white* in (**b**) and (**c**) and is overlayed with the incoming intensity distribution. The overall computation grid is 5 μm by 5 μm for all *panels*. The *panels* **d**, **e** and **f** show the corresponding normalized linear intensities in the far-field of the corresponding input field computed at roughly 4 Fresnel lengths behind the aperture. The wavelength for all experiments is 38 nm. One clearly sees that the phase distribution of the beams carrying a singularity causes additional, finer fringes. It can be expected that a reconstruction under the assumption of a known phase distribution of the impinging field would result in a better resolution in the cases of (**e**) and (**f**) compared to (**d**), where just a few smeared out fringes are visible

OV imaging has the potential to become a important topic in the coming years. Then immediately applicability in the XUV could be an outcome of the results presented in this work.

References

1. Curtis, J.E., Grier, D.G.: Structure of optical vortices. Phys. Rev. Lett. **90**(13), 133901 (2003)
2. Nye, J.F., Berry, M.V.: Dislocations in wave trains. Proc. R. Soc. Lond. A Math. **336**(1605), 165–190 (1974)
3. Soskin, M.S., Vasnetsov, M.V.: Singular optics. In: Wolf, E. (ed.) Progress in Optics, vol. 42, pp. 219–276. Elsevier, Amsterdam (2001)
4. Allen, L., Beijersbergen, M.W., Spreeuw, R.J., Woerdman, J.P.: Orbital angular momentum of light and the transformation of Laguerre-Gaussian laser modes. Phys. Rev. A **45**(11), 8185–8189 (1992)

5. Padgett, M.J., Allen, L.: The poynting vector in Laguerre-Gaussian laser modes. Opt. Commun. **121**(1–3), 36–40 (1995)
6. Simpson, N.B., Allen, L., Padgett, M. J.: Optical tweezers and optical spanners with Laguerre-Gaussian modes. J. Mod. Opt. **43**(12), 2485–2491 (1996)
7. Padgett, M., Bowman, R.: Tweezers with a twist. Nat. Photonics **5**(6), 343–348 (2011)
8. Allen, L., Padgett, M.J., Babiker, M.: IV The orbital angular momentum of light. In: Wolf, E. (ed.) Progress in Optics, vol. 39, pp. 291–372. Elsevier, New York (1999)
9. Beth, R.A.: Mechanical detection and measurement of the angular momentum of light. Phys. Rev. **50**(2), 115–125 (1936)
10. Volke-Sepulveda, K., Chavez-Cerda, S., Garces-Chavez, V., Dholakia, K.: Three-dimensional optical forces and transfer of orbital angular momentum from multiringed light beams to spherical microparticles. J. Opt. Soc. Am. B **21**(10), 1749–1757 (2004)
11. Grier, D.G.: A revolution in optical manipulation. Nature **424**(6950), 810–816 (2003)
12. Picon, A., Benseny, A., Mompart, J., de Aldana, J.R.V., Plaja, L., Calvo, G.F., Roso, L.: Transferring orbital and spin angular momenta of light to atoms. New J. Phys. **12**, 083053 (2010)
13. Molina-Terriza, G., Torres, J.P., Torner, L.: Twisted photons. Nat. Phys. **3**(5), 305–310 (2007)
14. Furhapter, S., Jesacher, A., Bernet, S., Ritsch-Marte, M.: Spiral interferometry. Opt. Lett. **30**(15), 1953–1955 (2005)
15. Oemrawsingh, S.S., van Houwelingen, J.A., Eliel, E.R., Woerdman, J.P., Verstegen, E.J., Kloosterboer, J.G., t Hooft, G.W.: Production and characterization of spiral phase plates for optical wavelengths. Appl. Opt. **43**(3), 688–694 (2004)
16. Swartzlander, G.A.: Achromatic optical vortex lens. Opt. Lett. **31**(13), 2042–2044 (2006)
17. Beijersbergen, M.W., Allen, L., Vanderveen, H.E., Woerdman, J.P.: Astigmatic laser mode converters and transfer of orbital angular-momentum. Opt. Commun. **96**(1–3), 123–132 (1993)
18. Heckenberg, N.R., McDuff, R., Smith, C.P., White, A.G.: Generation of optical phase singularities by computer-generated holograms. Opt. Lett. **17**(3), 221–223 (1992)
19. Bezuhanov, K., Dreischuh, A., Paulus, G.G., Schatzel, M.G., Walther, H.: Vortices in femtosecond laser fields. Opt. Lett. **29**(16), 1942–1944 (2004)
20. Peele, A.G., McMahon, P.J., Paterson, D., Tran, C.Q., Mancuso, A.P., Nugent, K.A., Hayes, J.P., Harvey, E., Lai, B., McNulty, I.: Observation of an x-ray vortex. Opt. Lett. **27**(20), 1752–1754 (2002)
21. Peele, A., Nugent, K.: X-ray vortex beams: a theoretical analysis. Opt. Express **11**(19), 2315–2322 (2003)
22. Cojoc, D., Kaulich, B., Carpentiero, A., Cabrini, S., Businaro, L., Di Fabrizio, E.: X-ray vortices with high topological charge. Microelectron. Eng. **83**(4–9), 1360–1363 (2006)
23. Firth, W.J., Skryabin, D.V.: Optical solitons carrying orbital angular momentum. Phys. Rev. Lett. **79**(13), 2450–2453 (1997)
24. Tikhonenko, V., Christou, J., Lutherdaves, B.: Spiraling bright spatial solitons formed by the breakup of an optical vortex in a saturable self-focusing medium. J. Opt. Soc. Am. B **12**(11), 2046–2052 (1995)
25. Vuong, L.T., Grow, T.D., Ishaaya, A., Gaeta, A.L., t Hooft, G.W., Eliel, E.R., Fibich, G.: Collapse of optical vortices. Phys. Rev. Lett. **96**(13), 133901 (2006)
26. Basistiy, I.V., Bazhenov, V.Y., Soskin, M.S., Vasnetsov, M.V.: Optics of light-beams with screw dislocations. Opt. Commun. **103**(5–6), 422–428 (1993)
27. Mamaev, A.V., Saffman, M., Zozulya, A.A.: Decay of high order optical vortices in anisotropic nonlinear optical media. Phys. Rev. Lett. **78**(11), 2108–2111 (1997)
28. Toda, Y., Honda, S., Morita, R.: Dynamics of a paired optical vortex generated by second-harmonic generation. Opt. Express **18**(17), 17796–17804 (2010)
29. Dreischuh, A., Paulus, G.G., Zacher, F., Grasbon, F., Neshev, D., Walther, H.: Modulational instability of multiple-charged optical vortex solitons under saturation of the nonlinearity. Phys. Rev. E **60**(6 Pt B), 7518–7524 (1999)
30. Henke, B.L., Gullikson, E.M., Davis, J.C: X-ray interactions: photoabsorption, scattering, transmission, and reflection at E=50-30000 eV, Z=1-92. At. Data Nucl Data. Tables **54** (2), 181–342 (1993)

31. Brabec, T., Krausz, F.: Intense few-cycle laser fields: frontiers of nonlinear optics. Rev. Mod. Phys. **72**(2), 545–591 (2000)
32. Basistiy, I.V., Soskin, M.S., Vasnetsov, M.V.: Optical wavefront dislocations and their properties. Opt. Commun. **119**(5–6), 604–612 (1995)
33. Soskin, M.S., Gorshkov, V.N., Vasnetsov, M.V., Malos, J.T., Heckenberg, N.R.: Topological charge and angular momentum of light beams carrying optical vortices. Phys. Rev. A **56**(5), 4064–4075 (1997)
34. Lohbreier, J., Eyring, S., Spitzenpfeil, R., Kern, C., Weger, M., Spielmann, C.: Maximizing the brilliance of high-order harmonics in a gas jet. New J. Phys. **11**(2), 023016 (2009)
35. Zuerch, M., Kern, C., Hansinger, P., Dreischuh, A., Spielmann, C.: Strong-field physics with singular light beams. Nat. Phys. **8**(10), 743–746 (2012)
36. Hernandez-Garcia, C., Picon, A., San Roman, J., Plaja, L.: Attosecond extreme ultraviolet vortices from high-order harmonic generation. Phys. Rev. Lett. **111**(8), 083602 (2013)
37. Neu, J.C.: Vortices in complex scalar fields. Physica D **43**(2–3), 385–406 (1990)
38. Neu, J.C.: Vortex dynamics of the nonlinear-wave equation. Physica D **43**(2–3), 407–420 (1990)
39. Fomichev, S.V.; Breger, P., Carre, B., Agostini, P., Zaretsky, D. F.: Non-collinear high-harmonic generation. Laser Phys. **12**(2), 383–388 (2002)
40. Eyring, S., Kern, C., Zuerch, M., Spielmann, C.: Improving high-order harmonic yield using wavefront-controlled ultrashort laser pulses. Opt. Express **20**(5), 5601–5606 (2012)
41. Lee, J.Y., Hong, B.H., Kim, W.Y., Min, S.K., Kim, Y., Jouravlev, M.V., Bose, R., Kim, K.S., Hwang, I.C., Kaufman, L.J., Wong, C.W., Kim, P., Kim, K.S.: Near-field focusing and magnification through self-assembled nanoscale spherical lenses. Nature **460**(7254), 498–501 (2009)
42. Synge, E.H.: A suggested method for extending microscopic resolution into the ultra-microscopic region. Philos. Mag. Ser. 7—XXXVIII. **6**(35), 356–362 (1928)
43. Hell, S.W.: Toward fluorescence nanoscopy. Nat. Biotechnol. **21**(11), 1347–1355 (2003)
44. Dan, D., Lei, M., Yao, B., Wang, W., Winterhalder, M., Zumbusch, A., Qi, Y., Xia, L., Yan, S., Yang, Y., Gao, P., Ye, T., Zhao, W.: DMD-based LED-illumination super-resolution and optical sectioning microscopy. Sci. Rep. **3**, 1116 (2013)
45. Volpe, G., Cherukulappurath, S., Juanola Parramon, R., Molina-Terriza, G., Quidant, R.: Controlling the optical near field of nanoantennas with spatial phase-shaped beams. Nano Lett. **9**(10), 3608–3611 (2009)
46. Tan, P.S., Yuan, X.C., Yuan, G.H., Wang, Q.: High-resolution wide-field standing-wave surface plasmon resonance fluorescence microscopy with optical vortices. Appl. Phys. Lett. **97**(24), 241109 (2010)
47. Dennis, M.R., Gotte, J.B.: Topological aberration of optical vortex beams: determining dielectric interfaces by optical singularity shifts. Phys. Rev. Lett. **109**(18), 183903 (2012)

Chapter 6
Summary and Outlook

The aim of this thesis was to implement microscopic imaging at the lab scale using coherent ultrafast XUV light sources. For this purpose two lensless imaging setups were built that both fit on a less than two by 1 m optical table.

The first setup that was built used a toroidal mirror and a grating to select the wavelength and to focus the XUV light onto the sample [1]. The advantage of this setup is that the operation wavelength can be freely chosen out of the harmonics spectrum and one can therefore cover a large range of wavelengths without breaking the vacuum. This makes this setup especially suited for phase contrast measurements. Moreover, it could be used in a multimodal operation combination with a spectrometer for detecting visible fluorescence of the sample. The disadvantage, however, is the relatively large focal spot size ($d \approx 50\,\mu\text{m}$ FWHM) induced by the HHG source size, which effectively limits the usable flux on a nanoscopic sample. For improving the latter a second setup that is based on dielectric mirrors for selecting the wavelength was built. By using two curved mirrors a tight focal spot size ($d \approx 3\,\mu\text{m}$ FWHM) was achieved, which allows to concentrate a high photon flux on the sample. The real advantage of the second setup, however, is that it was integrated into a new specifically built vacuum chamber that allows for high NA imaging in both transmission and reflection geometry.

A general drawback of CDI that was illustrated in this thesis is that usually several exposures must be combined in order to have a sufficient dynamic range in the diffraction pattern. Moreover, CDI is limited to isolated samples.[1] Digital in-line holography allows to overcome both limitations as it is demonstrated experimentally in this thesis. An effective algorithm that is inspired by iterative CDI methods was implemented in this work.[2] The DIH reconstruction algorithm proved to be very effective in reconstructing the holograms that were measured with the XUV HHG

[1] This is true at least for the CDI implementations presented in this thesis. In Sect. 2.5 novel less limited approaches reported in literature were discussed, e.g. ptychography.

[2] It should be noted that the general idea of using iterative CDI methods for holographic reconstructions is well-known [2]. However, the implementation presented in this thesis allows a reconstruction of complex-valued objects.

M.W. Zürch, *High-Resolution Extreme Ultraviolet Microscopy*,
Springer Theses, DOI 10.1007/978-3-319-12388-2_6

source. It was demonstrated that one can reconstruct digital in-line holograms almost in real-time, in contrast to the HIO algorithms used in CDI, which typically need a much longer time for a reconstruction. The samples that were investigated using DIH covered a wide range of possible applications. On a sample covered by a dirt layer even a three-dimensional profile of the sample could be extracted from a single measurement by using the Beer-Lambert law. At a destroyed silicon nitride membrane it was demonstrated that in principle the thickness of a phase-shifting layer can be resolved by fractions of the wavelength in a holographic reconstruction. Moreover, a sample that consists of nanoparticles that arranged like *diplococci* was imaged in great detail. On that sample a 900 nm polystyrene bead was resolved with a good contrast. Overall, it was demonstrated in this work that XUV DIH allows to access a sample even in three dimensions from a single hologram. It is worth mentioning that the samples that were imaged in this work, as far as could be ascertained, feature much more complexity compared to samples reported in literature in the field of XUV DIH.[3] DIH further offers a high experimental fidelity due to a relatively homogeneous illumination of the detector and, in contrast to CDI reconstructions, the mitigated requirements for noise-free diffraction data. However, it was also discussed that the size of the reference pinhole limits the achievable resolution to approximately the size of the pinhole. Hence, the practical limit of table-top XUV DIH would be in the range of a few 100 nm as long as pinholes are used as reference. The resolution achieved in the experiments presented in this thesis is slightly below 1 μm. Although the reconstruction is quick, the long exposure times that were several tens of minutes long in the experiments presented effectively limit the usability at this stage. A combination of the tight focusing dielectric mirror setup should allow for a quicker hologram acquisition in the near future.

On the other hand, the most widely used lensless imaging technique is coherent diffraction imaging. The advantage over DIH is that the resolution is only limited by the wavelength and the numerical aperture. Using the dielectric mirror setup, CDI measurements in transmission geometry at a high numerical aperture (NA = 0.8) achieved a resolution corresponding to 1.8 wavelengths by measuring a diffraction pattern of planar objects. To properly reconstruct the object from the diffraction pattern without any additional knowledge a curvature correction was applied. This accounts for the fact that one projects the Ewald sphere onto a planar detector. Further, an intensity normalization, originating from the fact the each pixel observes a different solid angle with respect to the aperture, was utilized. The measured diffraction pattern suggested a resolution of below one wavelength, which could, however, so far not be reproduced in the reconstruction. The reason found is that the planar projection of the data is limiting at this stage, owing to the fact that the sample has a finite depth and is actually a three dimensional object. Hence, for further experiments a full 3D reconstruction should be aimed for in order to take full advantage of the excellent beam properties produced by the fiber laser driven HHG source that was employed for these experiments. These were the first ever imaging experiments conducted with such a source. Moreover, the achieved resolution in terms of the

[3] Mostly opaque wires, needles or spider threads were imaged so far with XUV DIH.

wavelength is in the order of the best results that were ever achieved in table-top CDI experiments. The absolute resolution is just about 2.5 times lower than the world record.[4] The smaller size and better scalability of the FCPA lasers will certainly bring them ahead of the Ti:Sa based sources in the near future and probably even allow for real-time CDI with very short integration times while having a resolution in the order of the wavelength.

Furthermore, XUV CDI of a non-periodic sample in reflection geometry is demonstrated [4]. The HHG source was driven with a standard Ti:Sa laser. This was the first demonstration of reflection mode table-top XUV CDI with such a complexly shaped sample. For a high quality reconstruction, the data rescaling, such that the momentum transfer component q_x and q_y scale identically, was applied. Such rescaling allows the object to cover a minimal space in the reconstructed object plane, i.e. it appears not stretched. This reduces the support area needed and, hence, also reduces the effects induced by noisy data. A similar advantage of scaling the object 1:1 in the object plane was found elsewhere [5]. A spatial resolution of $\Delta r \approx 1.7\,\mu\text{m}$ was achieved. Theoretically, it was shown that for typical available wavelengths in reflection geometry one will likely be limited to a resolution of a few 100 nm, at least in the plane of the reflection, unless the angle of incidence is significantly increased, which then in turn will severely limit the reflected flux. Despite the resolution that was achieved in this work being on the order of those of standard light microscopes, it should be emphasized that the technique presented offers much more information about the object, e.g. three dimensional information from the object plane phase that is intrinsically retrieved. An improved resolution down to the limit mentioned above is straightforward and was only prevented due to mechanical limitations. Moreover, it was discussed that the reflection mode will likely be the geometry of choice for real-world applications of CDI. This is because samples can be much more easily prepared, e.g. by pipetting on highly reflecting surfaces in contrast to fragile ultra-thin membranes as they are essential for CDI in transmission geometry. Moreover, the outer shape of samples that highly absorb XUV, like biological specimens, can be probed in reflection geometry, because samples that have a high absorption typically feature good reflectivity at the same time. The most intriguing application of reflection mode CDI would, however, be microscopy of very fast processes on the surface. Since HHG sources can be naturally time-locked to the laser source, one could think of a kind of XUV pump-probe imaging, where a part of the HHG driving laser serves as the excitation pulse triggering dynamics in a sample. First results showing a promising temporal resolution are already reported [6, 7]. This would then result in a high resolution microscope in time and space.

A summary of the results achieved in this thesis in terms of imaging and in comparison to other published results is listed in Table 6.1. It can be concluded from the table that the experiments in this thesis compare very well to the best reported so far in terms of the relative resolution $\Delta r/\lambda$. The high NA experiment even reaches one of the best ever reported absolute resolutions for table-top systems

[4] Seaberg et al. [3] achieved $\Delta r \approx 22$ nm at $\lambda = 13.5$ nm. However, the sample had a less complex spatial structure.

Table 6.1 Summary of the achieved parameters with relevance to microscopy in contrast to results reported in literature

Source	Method	G	λ (nm)	NA	Δr (nm)	$\frac{\Delta r}{\lambda}$	FOV (μm)	References
HHG	CDI	T	33.2	0.8	60 (28)	1.8 (0.8)	~5	This work
HHG	CDI	T	13.5	0.79	22	1.6	~1	[3]
HHG	CDI	T	47	0.69	80	1.7	~4	[8]
HHG	CDI	T	32	0.57	62	1.9	~3	[9]
HHG	CDI	T	32	0.20	165	5.2	~20	[10]
SYN	CDI	T	0.89	0.01	150	169	~15	[11]
FEL	CDI	T	13.5	0.25	50	3.7	~5	[12]
FEL	CDI	T	32	0.26	62	1.9	~20	[13]
FEL	CDI	T	32.5	0.16	75	2.3	~10	[14]
HHG	CDI	R	38	0.03	1,700	45	~50	This work
HHG	CDI	R	29	0.29	$\gtrsim$100	~3.4	~3	[15]
SYN	CDI	R	0.17	0.02	22	129	~5	[5]
HHG	DIH	T	39	0.02	830	21	~100	This work
HHG	DIH	T	39	<0.01	6,800	174	~1,000	[16]
HHG	DIH	T	38	0.07	1,000	26	~100	[2]
SYN	FTH	T	1.59	0.04	50	31.4	~2	[17]
HHG	FTH	T	29	0.63	89	3.1	~2	[18]
HHG	FTH	T	32	0.56	110	3.4	~3	[19]

The table depicts different experimentally achieved parameters at different light sources (*SYN* synchrotron, *FEL* free-electron laser) in different geometries (G = T → transmission geometry, G = R → reflection geometry). The parameters are: the used wavelength λ, the numerical aperture NA, the resolution Δr, the relative resolution $\Delta r/\lambda$ and the field of view (*FOV*), which was estimated from reconstructions presented in the corresponding source. The table is sorted according to the method used: coherent diffraction imaging (*CDI*), digital in-line holography (*DIH*) and Fourier transform holography (*FTH*). There is a bias of the selected references towards HHG based experiments in order to compare those to the presented work. In the first row the achieved resolution suggested by the PRTF, as it is done in most papers, is denoted in brackets. Please note that this table does not claim completeness, especially in transmission CDI there is an enormous amount of papers available

with potential for improvement via including a 3D reconstruction. Additional to the techniques used in this work, one can find Fourier transform holography listed at the bottom. This technique uses nanometer scale references for producing the reference wave, which are specifically produced together with the sample. It shows that the resolution achievable with holographic methods can be almost identical to the best resolution achieved in CDI. Hence, it is beneficial to further improve digital in-line holography by reducing the reference size. The main limitation for table-top DIH will be to achieve a sufficient flux through the reference pinhole. Thus, a tight focusing of HHG beams as was presented for the dielectric mirror setup is essential.

Apart from imaging, the first demonstration that optical vortex beams can be generated at XUV wavelengths by direct conversion of an OV beam using HHG was given in this work. The experimental results indicate that the measured beam had a topological charge of one, in contradiction to theory and intuitive expectations for frequency up-conversion of OV beams. The behavior is explained by the imperfect beam shape in the focus, including some astigmatism, which results in the decay of the OV into singly charged vortices as it is expected from theory and found in many other experiments. These results were recently published [20]. In the scope of this thesis the application of XUV vortex beams to imaging was discussed. The result of a simulation was that in principle the radial phase distribution can be thought of as a kind of structured illumination that leads to sharper fringes in the far-field diffraction pattern of an object, which should allow for a reconstruction of objects beyond the Abbe limit. However, since a detailed analysis of this could fill a whole thesis, this simple first simulation shall be sufficient at this point.

In order to bring these lensless imaging techniques into an application, experiments on human breast cancer cells were conducted. The setup used was a mixture of DIH and CDI. It was demonstrated that one can distinguish two different breast cancer cell variations (MCF7 and SKBR3 cell line), i.e. cells of the same cancer type but with different expression profile and different genetic aberrations, directly from the measured diffraction pattern. The introduced method should allow for a clinical method to implement classification at a high throughput [21]. The idea behind the method is to treat the diffraction pattern, which is related to the outer shape of the object, as a kind of fingerprint of the object and compare it to a database of known objects. Surprisingly, it is already possible to distinguish between different cell expression profiles by using a very primitive picture comparison algorithm. One can expect that this method could be even further improved by implementing numerical picture comparison techniques known from e.g. face recognition. Again, the currently limiting factor for real-time analysis is the long exposure time ($t_{\mathrm{exp}} = 1,500\,\mathrm{s}$). Although the presented analysis was based on only four individual cells, this method should be further tested on a broader class of specimens in the near future. If successful, it could be used to identify cellular specimens on a daily basis with a high throughput without the need for additional preparation techniques such as labeling, staining or cultivation, as is necessary for comparable modern applied techniques in the field of clinical chemistry. A problem that was disregarded at this point is the need to bring the cells into the vacuum. In this thesis cells that were pipetted onto a substrate and dried. However, for a clinical application it would be beneficial to find a way to overcome this limitation. Very recently a possible way to image living cells in a vacuum environment was demonstrated [22]. Other fields of application for this technique could for instance be in microchip processing and micromachining where deviations of the shape of the object at the sub-micron scale could be directly identified from the diffraction pattern.

References

1. Zuerch, M., Spielmann, C.: Patent: Verfahren und Vorrichtung zur Erzeugung einer schmalbandigen, kurzwelligen, kohaerenten Laserstrahlung, insbesondere fuer XUV-Mikroskopie. DE102012022961.5 - int. reg. PCT/DE2013/000695, 21 Nov 2012 (2012)
2. Genoud, G., Guilbaud, O., Mengotti, E., Pettersson, S.G., Georgiadou, E., Pourtal, E., Wahlstroem, C.G., L'Huillier, A.: XUV digital in-line holography using high-order harmonics. Appl. Phys. B **90**(3–4), 533–538 (2008)
3. Seaberg, M.D., Adams, D.E., Townsend, E.L., Raymondson, D.A., Schlotter, W.F., Liu, Y.W., Menoni, C.S., Rong, L., Chen, C.C., Miao, J.W., Kapteyn, H.C., Murnane, M.M.: Ultrahigh 22 nm resolution coherent diffractive imaging using a desktop 13 nm high harmonic source. Opt. Express **19**(23), 22470–22479 (2011)
4. Zuerch, M., Kern, C., Spielmann, C.: XUV coherent diffraction imaging in reflection geometry with low numerical aperture. Opt. Express **21**(18), 21131–21147 (2013)
5. Sun, T., Jiang, Z., Strzalka, J., Ocola, L., Wang, J.: Three-dimensional coherent X-ray surface scattering imaging near total external reflection. Nat. Photonics **6**(9), 588–592 (2012)
6. Reis, D.A., DeCamp, M.F., Bucksbaum, P.H., Clarke, R., Dufresne, E., Hertlein, M., Merlin, R., Falcone, R., Kapteyn, H., Murnane, M.M., Larsson, J., Missalla, T., Wark, J.S.: Probing impulsive strain propagation with X-ray pulses. Phys. Rev. Lett. **86**(14), 3072–3075 (2001)
7. Tobey, R.I., Siemens, M.E., Cohen, O., Murnane, M.M., Kapteyn, H.C., Nelson, K.A.: Ultrafast extreme ultraviolet holography: dynamic monitoring of surface deformation. Opt. Lett. **32**(3), 286–288 (2007)
8. Raines, K.S., Salha, S., Sandberg, R.L., Jiang, H., Rodriguez, J.A., Fahimian, B.P., Kapteyn, H.C., Du, J., Miao, J.: Three-dimensional structure determination from a single view. Nature **463**(7278), 214–217 (2010)
9. Ravasio, A., Gauthier, D., Maia, F.R., Billon, M., Caumes, J.P., Garzella, D., Geleoc, M., Gobert, O., Hergott, J.F., Pena, A.M., Perez, H., Carre, B., Bourhis, E., Gierak, J., Madouri, A., Mailly, D., Schiedt, B., Fajardo, M., Gautier, J., Zeitoun, P., Bucksbaum, P.H., Hajdu, J., Merdji, H.: Single-shot diffractive imaging with a table-top femtosecond soft X-ray laser-harmonics source. Phys. Rev. Lett. **103**(2), 028104 (2009)
10. Chen, B., Dilanian, R.A., Teichmann, S., Abbey, B., Peele, A.G., Williams, G.J., Hannaford, P., Van Dao, L., Quiney, H.M., Nugent, K.A.: Multiple wavelength diffractive imaging. Phys. Rev. A **79**(2), 023809 (2009)
11. Abbey, B., Whitehead, L.W., Quiney, H.M., Vine, D.J., Cadenazzi, G.A., Henderson, C.A., Nugent, K.A., Balaur, E., Putkunz, C.T., Peele, A.G., Williams, G.J., McNulty, I.: Lensless imaging using broadband X-ray sources. Nat. Photonics **5**(7), 420–424 (2011)
12. Barty, A., Boutet, S., Bogan, M.J., Hau-Riege, S., Marchesini, S., Sokolowski-Tinten, K., Stojanovic, N., Tobey, R., Ehrke, H., Cavalleri, A., Dãœsterer, S., Frank, M., Bajt, S., Woods, B.W., Seibert, M.M., Hajdu, J., Treusch, R., Chapman, H.N.: Ultrafast single-shot diffraction imaging of nanoscale dynamics. Nat. Photonics **2**(7), 415–419 (2008)
13. Chapman, H.N., Barty, A., Bogan, M.J., Boutet, S., Frank, M., Hau-Riege, S.P., Marchesini, S., Woods, B.W., Bajt, S., Benner, H., London, R.A., Plonjes, E., Kuhlmann, M., Treusch, R., Dusterer, S., Tschentscher, T., Schneider, J.R., Spiller, E., Moller, T., Bostedt, C., Hoener, M., Shapiro, D.A., Hodgson, K.O., Van der Spoel, D., Burmeister, F., Bergh, M., Caleman, C., Huldt, G., Seibert, M.M., Maia, F.R.N.C., Lee, R.W., Szoke, A., Timneanu, N., Hajdu, J.: Femtosecond diffractive imaging with a soft-X-ray free-electron laser. Nat. Phys. **2**(12), 839–843 (2006)
14. Capotondi, F., Pedersoli, E., Mahne, N., Menk, R.H., Passos, G., Raimondi, L., Svetina, C., Sandrin, G., Zangrando, M., Kiskinova, M., Bajt, S., Barthelmess, M., Fleckenstein, H., Chapman, H.N., Schulz, J., Bach, J., Froemter, R., Schleitzer, S., Mueller, L., Gutt, C., Gruebel, G.: Invited article: coherent imaging using seeded free-electron laser pulses with variable polarization: first results and research opportunities. Rev. Sci. Instrum. **84**(5), 051301 (2013)

15. Gardner, D.F., Zhang, B., Seaberg, M.D., Martin, L.S., Adams, D.E., Salmassi, F., Gullikson, E., Kapteyn, H., Murnane, M.: High numerical aperture reflection mode coherent diffraction microscopy using off-axis apertured illumination. Opt. Express **20**(17), 19050–19059 (2012)
16. Bartels, R.A., Paul, A., Green, H., Kapteyn, H.C., Murnane, M.M., Backus, S., Christov, I.P., Liu, Y.W., Attwood, D., Jacobsen, C.: Generation of spatially coherent light at extreme ultraviolet wavelengths. Science **297**(5580), 376–378 (2002)
17. Eisebitt, S., Luning, J., Schlotter, W.F., Lorgen, M., Hellwig, O., Eberhardt, W., Stohr, J.: Lensless imaging of magnetic nanostructures by X-ray spectro-holography. Nature **432**(7019), 885–888 (2004)
18. Sandberg, R.L., Raymondson, D.A., La, O.V.C., Paul, A., Raines, K.S., Miao, J., Murnane, M.M., Kapteyn, H.C., Schlotter, W.F.: Tabletop soft-X-ray Fourier transform holography with 50 nm resolution. Opt. Lett. **34**(11), 1618–1620 (2009)
19. Gauthier, D., Guizar-Sicairos, M., Ge, X., Boutu, W., Carre, B., Fienup, J.R., Merdji, H.: Single-shot femtosecond X-ray holography using extended references. Phys. Rev. Lett. **105**(9), 093901 (2010)
20. Zuerch, M., Kern, C., Hansinger, P., Dreischuh, A., Spielmann, C.: Strong-field physics with singular light beams. Nat. Phys. **8**(10), 743–746 (2012)
21. Zuerch, M., Spielmann, C.: Verfahren zur Auswertung von durch schmalbandige, kurzwellige, kohaerente Laserstrahlung erzeugten Streubildern von Objekten, insbesondere zur Verwendung in der XUV-Mikroskopie. DE102012022966.6 - int. reg. PCT/DE2013/000696, 21 Nov 2012 (2012)
22. Kimura, T., Joti, Y., Shibuya, A., Song, C., Kim, S., Tono, K., Yabashi, M., Tamakoshi, M., Moriya, T., Oshima, T., Ishikawa, T., Bessho, Y., Nishino, Y.: Imaging live cell in micro-liquid enclosure by X-ray laser diffraction. Nat. Commun. **5**, 3052 (2014)

Appendix A
XUV Mirror Ray Trace and Aberration Minimized Setup

In Sect. 3.2.2 the CDI setup based on dielectric mirrors for selecting the wavelength was described. Moreover, the design goal was to achieve a tighter focal spot by first collimating the beam and focusing it down by a curved mirror having a short focal length. With the grating based setup (Sect. 3.2.1) a focal spot size of roughly 50 μm (FWHM) was achieved, which corresponds to a one to one imaging of the source by means of a $4f$-setup. For analyzing the dielectric mirror setup, the powerful ray-tracing software ZEMAX was employed to investigate numerically the aberrations of the setup. What is clear beforehand is that one expects primarily astigmatism from the focusing mirrors, which increases as the angle of incidence gets larger. As a source for the ray trace, rays homogeneously emitted from a disk having a diameter of 50 μm were initiated. The divergence is set such that the first mirror is illuminated with a beam having a diameter of roughly 10 mm which compares well with the experimental conditions in this work.

The setup consists of a standard z-fold (Fig. A.1a) using two focusing mirrors. M1 has a focal length of $f_1 = 1{,}000$ mm and M2 has a focal length of $f_2 = 500$ mm. The first mirror is placed f_1 away from the source. The distance between M1 and M2 is maximized but constrained by the chamber to about ~200 mm. Since the mirrors consist of standard 1 inch sized substrates and the XUV beam has a diameter of roughly 10 mm on M1 there is a restriction on the amount of quenching the z-fold. An angle of incidence[1] of 5° to the normal of M1 and M2 was chosen (Fig. A.1a). The result was that the beam towards M1 by M2 was not obscured and low astigmatism

[1] As an exception in this appendix section the angle of incidence is measured against the surface normal to stay consistent with ZEMAX.

M.W. Zürch, *High-Resolution Extreme Ultraviolet Microscopy*,
Springer Theses, DOI 10.1007/978-3-319-12388-2

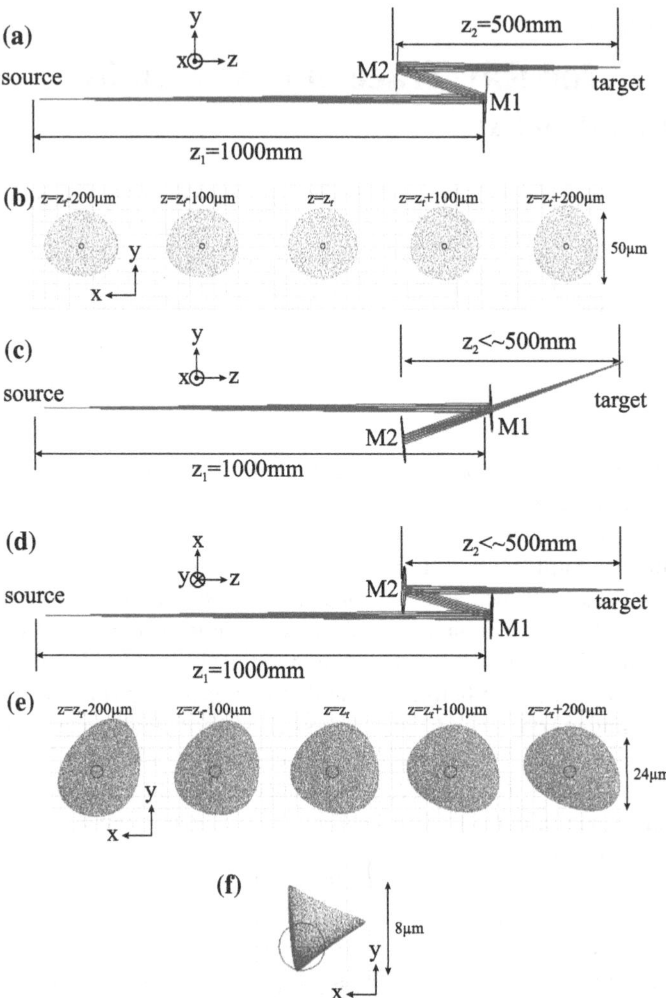

Fig. A.1 Different layouts for the CDI beam focusing setup. **a** *Top view* Layout as it was in principle built for the high NA CDI experiments (Sect. 4.2). The two focusing mirrors M1 and M2 are deflecting both in the y-z-plane. M1 collimates the beam and M2 focuses the beam down onto the target, i.e. the sample to be examined using CDI. **b** Ray trace results of rays starting equally distributed on a disk of 50 μm diameter in the source. The result consists of 5 parallel x-y-planes scanned in 100 μm steps around the focus z_f. **c** *Top view* Optimized setup for better focusing. M1 directs the beam out of y-z-plane onto M2 which is vertically displaced, see **d** *side view*, which brings the beam parallel to the y-z-plane again. **e** Shows the corresponding ray trace result which features a two times tighter focal spot compared to (**b**), see text for an explanation. **f** Allowing ZEMAX to choose the exit angles at M2 freely, one can even achieve an almost diffraction limited focus. However, having a residual angle subtending the y-z-plane is not feasible for CDI experiments unless both the camera, which is typically mounted parallel to the chamber, and the sample can be tilted freely to again guarantee normal incidence on the sample. Please note that panels **a**, **c** and **d** are to scale 1:2 in the z-axis

was achieved. Both mirrors keep the light parallel to the y-z-plane. The ray trace results in a scan around the focus (Fig. A.1b) revealing rays arriving up to a radial diameter of 50 μm at maximum. The shape is almost perfectly round, but a small astigmatism is already visible. For angles of incidence significantly larger than 5° the quality of the focus degrades quickly. The Airy disc, i.e. the region where the rays would hit in absence of any aberration (≡M1–M2 on-axis), is about 2 μm in diameter. One should note that a ray trace does not enable the determination of the expected FWHM diameter of a real beam. This size was determined experimentally by scanning a 2 μm pinhole across the focus. It was found to be in the range of 5 μm, which is also confirmed by high NA CDI measurements of objects being a little larger than 5 μm in diameter.[2] These objects lacked intensity in the reconstruction at the edges due to the missing photon flux.

The problem as to why the z-fold in Fig. A.1a does not reach the diffraction limit (illustrated by the Airy disc) is because the astigmatism introduced by M1 adds to the astigmatism of M2 in the same axis. Hence the overall astigmatism of the system is not optimal. To improve this, the setup depicted in Fig. A.1c, d is presented. M1 now is hit again under angle of incidence of 5° but now it deflects the beam out of the y-z-plane such that M2 is hit under the following angles of incidence with respect to the subscripted planes: $\alpha_{yz} = 0.4°$ and $\alpha_{xz} = 4°$. This results in a rotation of the system around the z-axis. Thus M2 contributes its astigmatism in the perpendicular cross section and hence the overall astigmatism is smaller compared to the previous setup. The astigmatisms do not fully cancel each other out, because intrinsically the wavefront is different on each mirror aside from different radii of curvature. The ray trace is depicted in Fig. A.1d. One can see that the radius of the ray scatter in the focus is two times smaller compared to Fig. A.1b. This means practically that one can focus the beam better by simply bringing one mirror out of plane. As a constraint for the data presented in Fig. A.1c–e the beam was forced to come out of the system parallel to the y-z plane, i.e. the optical table. If ZEMAX is allowed to optimize the exit angle of M2 in such a way that the outgoing beam is allowed to have a residual angle of $<5°$ subtending the y-z-plane, one gets the result depicted in Fig. A.1f. It shows that the rays almost all hit within the Airy disk. The triangle structure reveals that only residual coma aberration is left in that arrangement with a beam diameter almost only limited by diffraction.

The setup presented in this appendix (Fig. A.2) could be readily used for future setups where an even tighter focus may be of use. In terms of a Gaussian focus the focus physically present in the case of the geometry used to generate Fig. A.1f should

[2] The deviation of a factor of 10 for the ray trace is still not explained. It is likely that the source size of the FCPA laser driven HHG source is much smaller due to the harder focusing compared to the Ti:Sa driven HHG source for which the ray trace was computed.

be around 2 μm (FWHM), which is the diffraction limit given by the NA of M2.

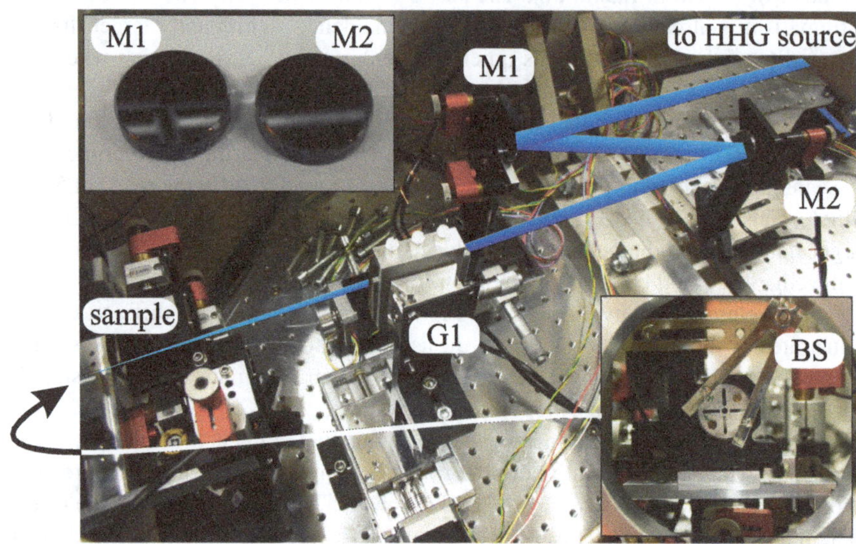

Fig. A.2 Photographic view of the dielectric mirror based CDI chamber. The XUV HHG source (Fig. 3.2) emits from the *right*. Two dielectric coated mirrors (detailed on the *inset* in the *left upper corner*) M1 and M2 refocus the light onto the sample for CDI. A grating G1 can be introduced to the beam path in grazing incidence in order to check the spectrum reflected by the mirrors. The sample area is detailed on the *inset* in the *lower right corner* as seen from the CCD which would follow to the *left* in the overview picture. The sample is mounted on a aluminum mount having 4 long slits which are used for alignment of the beam before the sample is driven into the beam. The beam stop (BS) consisting of a sugar crystal ($d \approx 100\,\mu\text{m}$) fixed to a 5 μm tungsten wire (too tiny to be seen in the photograph) is affixed to a fork-like construction that can be independently driven into the beam path. All optics and mounts can be electrically driven using piezo motors having a step size $\approx 30\,\frac{\text{nm}}{\text{step}}$

Appendix B
Simulations of Fringe Patterns for Different Topological Charges in XUV Vortex Beams

In Sect. 5.3 the fringe pattern of an XUV vortex beam caused by a thin wire observed experimentally was compared to a simulation. Here the simulation will be described. Figure B.1a shows the intensity distribution of the numerically generated $\text{TC} = 1$ vortex beam overlayed with a thin wire that is set to split the wavefront. The fringe pattern in the far-field is calculated by using the numerical methods described in Sect. 2.4. The field obscured by the wire is propagated numerically by approximately 4 Fresnel lengths to be safely in the far-field. The simulation of the fringe pattern was carried out for several input TCs: $1, 3, 5, 23$. These correspond mainly to the expected TCs for low-order harmonics and the high one for the one expected for the 23rd harmonic. The results are depicted in Fig. B.1b–e. One clearly sees that such fringe patterns are easily distinguishable by their shape. Note the decreasing width of the ring for higher TCs. For very high topological charges the fringe pattern appears completely spoiled, which is due to the already quite large phase shift from one horizontal end of the wire to the other. The simulation for the high TC was carried out on a $6{,}000$ by $6{,}000$ pixel grid to assure that the phase is sampled densely enough. However, it is clear from the depicted simulation results that very high topological charges can be ruled out in the presented experiment. For the low charges the $\text{TC} = 1$ appears most likely, because it also corresponds to the approximate relation of ring width and dark core width in the measured data. Unfortunately, it was not possible to perform a more accurate simulation by including all experimental parameters, for example by calculating the size of the vortex mode at the wire and thus find the correct relation between wire width and vortex width. The reason for this is mainly that the exact source size is only very imprecisely known and thus the Rayleigh length cannot be calculated sufficiently accurately. This is further complicated, because the high harmonics are generated in a gas cell slightly behind the focus of the IR beam to maintain phase matching due to the Gouy-phase. Hence, the exact divergence of the HHG beam and its spatial size cannot be determined accurately enough. Thus only a qualitative look is taken at the far-field fringe pattern of an OV obscured by a

M.W. Zürch, *High-Resolution Extreme Ultraviolet Microscopy*,
Springer Theses, DOI 10.1007/978-3-319-12388-2

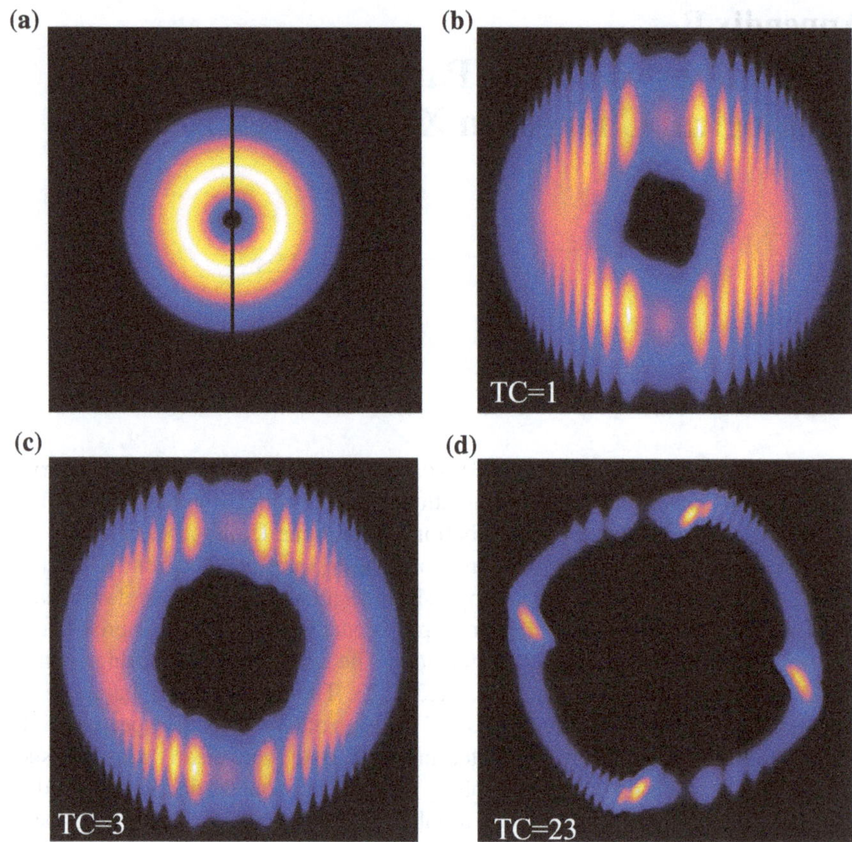

Fig. B.1 Simulation of fringe patterns of optical vortices carrying different TCs at a thin opaque wire. **a** The numerically generated OV with TC = 1 overlayed with a thin opaque wire. Far-field fringe pattern after propagating 4 Fresnel lengths for **b** TC = 1, **c** TC = 3 and **d** TC = 23. Note the small increase of the mode diameter after propagating the distance. All panels show the normalized intensity. Panels **b**–**d** correspond to the same distance behind the interaction region

thin wire. For further experiments it would be beneficial to either scan the wire along the XUV beam propagation axis and record several fringe patterns or to transverse the wire across the XUV beam. From either of these measurements more accurate simulations could be carried out.

Curriculum Vitae

Personal Data

Name: Michael Werner Zürch
Date of Birth: 16.03.1986
Pace of Birth: Werdau
Nationality: German

Education

1992–1996 Elementary school at Grundschule Trünzig.
1996–2002 Secondary school at Diesterweg Mittelschule Werdau, with distinction (average 1.3).
2002–2005 Abitur at Technisches Gymnasium Werdau, with distinction (average 1.0).

M.W. Zürch, *High-Resolution Extreme Ultraviolet Microscopy*,
Springer Theses, DOI 10.1007/978-3-319-12388-2

Scientific Education

2005–2010 Diploma studies in technical physics at Friedrich-Schiller-University Jena, finished with distinction (average 1.1) with the topic of the diploma thesis *Limitations of Ultra-Fast Nonlinear Optics in Nanostructured Samples*.

2008 Student research project at Institute for Solid State Physics, FSU Jena, with the topic *Noise measurement on HTSL-gradiometers fabricated on crossed bi-crystals.*

2010–2014 Doctorate in physics at Institute of Optics and Quantumelectronics and Abbe Center of Photonics, FSU Jena, finished with distinction (*summa cum laude*) on the topic *Coherent High-Resolution Imaging of Artificial and Biological Specimens using Compact Ultrafast Extreme Ultraviolet Sources.*

Employment Career

2009 Undergraduate assistant at Institute for Optics and Quantumelectronics, FSU Jena.

Since 2010 Scientific assistant at Institute for Optics and Quantumelectronics, FSU Jena.

Scientific Awards

- Best diploma thesis award of the Faculty of Physics and Astronomy, FSU Jena, in 2010.
- Student paper award first place in Physics of Medical Imaging, SPIE Medical Imaging Conference, San Diego, USA, in 2014.

Language Skills

- German (native)
- English (language proficient)
- Russian (A2 level)
- Danish (A2 level)

Teaching Qualification

- Advisor for lab works for undergraduate students over four years.
- Seminar supervisor for the seminar *Fundamentals of Photonics* over four years.
- Co-Supervisor for several Bachelor and Master Thesis.
- Participation in the full-year *Teaching Qualification Basic* course of the Graduate Academy at FSU Jena.

Articles in Peer-reviewed Journals

1. **M. Zürch**, M. Kanka, R. Riesenberg, C. Spielmann. Extreme ultraviolet digital in-line holography using a table top source, submitted.

2. **M. Zürch**, J. Rothhardt, S. Hädrich, S. Demmler, M. Krebs, J. Limpert, A. Tünnermann, A. Guggenmos, U. Kleineberg, and Ch. Spielmann. Real-time and Sub-wavelength Ultrafast Coherent Diffraction Imaging in the Extreme Ultraviolet, submitted.
3. **M. Zürch**, S. Foertsch, M. Matzas, K. Pachmann, R. Kuth, and Ch. Spielmann. Fast method for cellular classification based on coherent high resolution XUV diffraction patterns, *Journal of Medical Imaging* **1** (3), 031008 (2014).
4. A. Hoffmann, **M. Zürch**, M. Gräfe, and Ch. Spielmann. Spectral broadening and compression of sub-millijoule laser pulses in hollow-core fibers filled with sulfur hexafluoride, *Optics Express* **22**, 12038–12045 (2014).
5. **M. Zürch**, A. Hoffmann, M. Gräfe, B. Landgraf, M. Riediger, and Ch. Spielmann. Characterization of a broadband interferometric autocorrelator for visible light with ultrashort blue laser pulses, *Optics Communications* **321**, 28–31 (2014).
6. **M. Zürch**, C. Kern, and Ch. Spielmann. XUV coherent diffraction imaging in reflection geometry with low numerical aperture, *Optics Express* **21** (18), 21131–21147 (2013).
7. **M. Zürch**, C. Kern, P. Hansinger, A. Dreischuh, and Ch. Spielmann. Strong-field physics with singular light beams, *Nature Physics* **8**, 743–746 (2012).
8. S. Eyring, C. Kern, **M. Zürch**, and Ch. Spielmann. Improving high-order harmonic yield using wavefront-controlled ultrashort laser pulses, *Optics Express* **20** (5), 5601–5606 (2012).
9. C. Kern, **M. Zürch**, J. Petschulat, T. Pertsch, B. Kley, T. Käsebier, U. Hübner, and Ch. Spielmann. Comparison of femtosecond laser-induced damage on unstructured versus nano-structured Au-targets, *Applied Physics A* **104** (1), 15–21 (2011).

Articles in Proceedings

1. **M. Zürch**, S. Foertsch, M. Matzas, K. Pachmann, R. Kuth, and Ch. Spielmann. Apparatus and fast method for cancer cell classification based on high harmonic coherent diffraction imaging in reflection geometry, *Proceedings of SPIE—The International Society for Optical Engineering* **9033**, Medical Imaging 2014: Physics of Medical Imaging, Art. No. 9033–58 (2014).
2. Ch. Spielmann, **M. Zürch**, C. Kern, P. Hansinger, and A. Dreischuh. Extreme nonlinear optical processes with beams carrying orbital angular momentum, *Proceedings of SPIE—The International Society for Optical Engineering* **8984**, Ultrafast Phenomena and Nanophotonics XVIII, Art. No. 8984–45 (2014).

Book

M. Zürch, *High-Resolution Extreme Ultraviolet Microscopy: Imaging of Artificial and Biologic Specimens with Laser-driven Ultrafast XUV Sources* published within Springer Theses, Springer Berlin Heidelberg, 2015.

Patents

1. **M. Zürch** and Ch. Spielmann, *Verfahren zur Auswertung von durch schmalbandige, kurzwellige, kohärente Laserstrahlung erzeugten Streubildern von*

Objekten, insbesondere zur Verwendung in der XUV-Mikroskopie, patent pending DE102012022966.6, int. reg. PCT/DE2013/000696, 21.11.12.

2. **M. Zürch** and Ch. Spielmann,*Verfahren und Vorrichtung zur Erzeugung einer schmalbandigen, kurzwelligen, kohärenten Laserstrahlung, insbesondere für XUV-Mikroskopie*, patent pending DE102012022961.5, int. reg. PCT/DE2013/000695, 21.11.12.

Conference Contributions

1. **M. Zürch**, J. Rothhardt, S. Hädrich, S. Demmler, M. Krebs, J. Limpert, A. Tünnermann, A. Guggenmos, U. Kleineberg, and Ch. Spielmann. High Average Power quasi-monochromatic XUV Source for Coherent Imaging Applications, *High-Intensity Lasers and High-Field Phenomena*, Berlin, Germany, 20.03.2014. (talk)
2. **M. Zürch**, S. Foertsch, M. Matzas, K. Pachmann, R. Kuth, and Ch. Spielmann. Apparatus and fast method for cancer cell classification based on high harmonic coherent diffraction imaging in reflection geometry, *SPIE Medical Imaging 2014*, San Diego, USA, 20.02.2014. (talk)
3. Ch. Spielmann, **M. Zürch**, C. Kern, P. Hansinger, and A. Dreischuh. Extreme nonlinear optical processes with beams carrying orbital angular momentum, *SPIE Photonics West—Ultrafast Phenomena and Nanophotonics XVIII*, San Francisco, USA, 05.02.2014. (invited talk)
4. **M. Zürch**, S. Foertsch, M. Matzas, K. Pachmann, R. Kuth, and Ch. Spielmann. Compact Coherent High Resolution XUV Microscopy Device for Imaging of Biologic Specimen, *American Association for Clinical Chemistry Annual Meeting 2013*, Houston, USA, 31.07.2013. (poster)
5. **M. Zürch**, S. Foertsch, M. Matzas, K. Pachmann, R. Kuth, and Ch. Spielmann. Fast Method for Classification Based on Coherent High Resolution XUV Scatter Images of Biologic specimen, *American Association for Clinical Chemistry Annual Meeting 2013*, Houston, USA, 30.07.2013. (poster)
6. C. Kern, **M. Zürch**, P. Hansinger, A. Dreischuh, and Ch. Spielmann. Extreme Nonlinear Optical Processes with Beams Carrying Orbital Angular momentum, *CLEO/Europe—IQEC Conference*, Munich, Germany, 16.05.2013. (talk)
7. **M. Zürch**, C. Kern, and Ch. Spielmann. Tabletop Lensless Imaging Apparatus using an Ultrashort High Harmonic XUV Source, *CLEO/Europe—IQEC Conference*, Munich, 16.05.2013. (poster)
8. **M. Zürch**, C. Kern, P. Hansinger, A. Dreischuh, and Ch. Spielmann. Extreme nonlinear optics with singular light beams carrying orbital angular momentum, *Ultrafast Optics Conference*, Davos, Switzerland, 08.03.2013. (talk)
9. C. Kern, **M. Zürch**, B. Kley, T. Pertsch, and Ch. Spielmann. Considerations and Requirements of Metallic Nanostructures for Plasmon-Enhanced High-Harmonic Generation, *Ultrafast Optics Conference*, Davos, Switzerland, 05.03.2013. (poster)
10. **M. Zürch**, C. Kern, P. Hansinger, A. Dreischuh, and Ch. Spielmann. Highly nonlinear interaction of laser beams with orbital angular momentum, *DokDok 2012 Conference*, Oppurg 2012, 7–11.10.2012. (poster)

11. C. Kern, **M. Zürch**, J. Petschulat, and Ch. Spielmann. Limitations of extreme nonlinear ultrafast nano-photonics, *Nano and Photonics Conference*, Mauterndorf 2011, 18.03.2011. (poster)
12. **M. Zürch**, C. Kern, J. Petschulat, T. Pertsch, B. Kley, T. Käsebier, U. Hübner, and Ch. Spielmann. Femtosecond Damage Threshold of Plasmonic Nanostructures Compared to Au Films, *4th European Conference on Applications of Femtosecond Lasers in Materials Science—FemtoMat 2011*, Mauterndorf, Austria, 16.03.2011. (talk)
13. C. Kern, **M. Zürch**, and Ch. Spielmann. Limitations of ultrafast nonlinear nano-optics, *International Summer School in Ultrafast Nonlinear Optics 2010*, Heriot-Watt University, Edinburgh, 11.08.2010. (poster)

Articles in Non-peer-reviewed Journals

1. **M. Zürch**, and Ch. Spielmann. Diagnose von Krebszellen mittels Ultrakurzpulslaser, *BioPhotonik* **7** (1), 36–39 (2014).
2. **M. Zürch**, C. Kern, and Ch. Spielmann. Optische Wirbel, *Physik in unserer Zeit* **3** (44), 134–141 (2013).
3. **M. Zürch**. Die Jenaer Physiker und der Amateurfunk, *CQ-DL das Amateurfunkmagazin* **11**, 805–807 (2011).

Zeitfracht Medien GmbH
Ferdinand-Jühlke-Straße 7
99095 Erfurt, Deutschland
produktsicherheit@kolibri360.de